The Magic of
the Collective

The Magic of the Collective

A Life in the Service of Science

Kai Simons

Copyright © 2023

All Rights Reserved

No part of this book may be reproduced or transmitted in any form or by any means, electronic or mechanical, including photocopying, recording, or by any information storage and retrieval system without the written permission of the author, except where permitted by law.

ISBN: 978-1-917007-02-3

*For Carola, my love, companion and partner,
and our beloved children, Katja, Mikael and Matias*

Table of Contents

Prologue i

Chapter 1: Helsinki 1

Chapter 2: New York 22

Chapter 3: Back in Helsinki 31

Chapter 4: Heidelberg 53

Chapter 5: Dresden 129

Epilogue 222

Scientists Working in the Simons Lab 228

Acknowledgements 230

About the Author 233

Prologue

This is a book about the ups and downs of doing research. Early on, I became intrigued by the power of cooperation. Why do we so often pursue only our self-interest instead of teaming up with others? Why don't we collaborate more? After moving into research, it became even more obvious to me that I needed to work together with other scientists to be able to solve the research problems that I faced.

Collaboration is not the guiding principle for how bio-scientists usually work. We are brought up to function as individuals. The loneliness of the pursuit is driving many of us to despair. Even if we work in groups, it's individual success that counts. This has led to working environments that are dominated by egos and competition. To move on in your research career, you must focus on 'what's in it for me' and constantly market yourself; otherwise, it's difficult to become an independent principal investigator and be promoted to tenure. This overly competitive environment is one factor that scares away less 'pushy' scientists, including many women. This is a lost opportunity because their talent is badly needed to balance our research community.

The scientific process requires a collective spirit to move innovations into practice. Science is self-correcting. Errors and misinterpretations are corrected when novel insights emerge. This provides a dynamic and important driving force for change in our society that allows our body of knowledge to be constantly corrected. However, this process must be nurtured. Can we remain intellectually humble and acknowledge that we might be wrong? We scientists have often failed at this, and we continue to fail. We are blinded by the success of science and technology, which has generated untold benefits for mankind, prolonging our life expectancy and improving our quality of life. We forget that scientific breakthroughs often generate serious new problems. This has become especially clear as we see that the technologies that increase our global welfare have also been instrumental in increasing the CO_2 emissions that are fuelling the climate disaster.

We saw this dichotomy between the positive effects and possible negative consequences of innovations play out again in the recent COVID-19 pandemic. We witnessed a wave of scientific collaboration prompted by the onslaught of the pandemic. The RNA genome of the virus SARS-Cov-2 was sequenced rapidly. This sequence provided the foundation for research that, in no time, led to effective vaccines against the virus infection. This global investment succeeded beyond expectations. But it soon became clear that the success of the vaccines was not sufficient to stop the pandemic; they first had to be distributed, and people had to be convinced to take them. This was a much more demanding challenge than we were prepared for. The counterwave of distrust that the scientific success encountered was not a consequence of the generation of effective vaccines but probably reflected a general scepticism of science that has spread through our communities after previous experiences with scientific breakthroughs.

The importance of science in society is steadily increasing, and this demands an increased responsibility on the part of scientists for how we innovate. Basic science feeds technology, and thus scientists and engineers are in the same boat. Together, we must learn to navigate the hidden rocks that might cause us to capsize. Science provides an immensely powerful force to change life on our planet for good. However, there are so many unintended consequences of scientific discoveries. We are damned to continue to innovate just to correct problems that previous generations of scientists have generated. Our focus has been too narrow. We need to learn to cooperate with other research areas to succeed in producing solutions that are sustainable.

We need better practices for doing research collectively. Promoting cooperation has been the leitmotif of my life. My research career took off in Helsinki, which was far from the centres of excellence that dominated the molecular life sciences then. This perspective from afar opened my eyes and led me to search with tenacity for new solutions to create cooperative research habitats in competitive environments. One driving reason for writing this book is to describe how to build functional and viable research environments based on cooperation and partnership. My colleagues and I did this in Helsinki in the newly founded Haartman Institute of the medical faculty, in Heidelberg, where the European Molecular Biology Laboratory started its operation in 1974 under the leadership of Sir John Kendrew, and, finally, in Dresden, where we built the Max Planck Institute of Molecular Cell Biology and Genetics. With this book, I want to share the lessons I learned in this lifelong process. It is guided by the belief that more altruism in the practice of science will help us not only to improve our service to society but also to generate scientific environments that are more welcoming to all scientists.

Chapter 1: Helsinki

I was sixteen, hanging out with my friends on our bikes, wondering how to pass the time. Across the street, there was a group of girls, but we were all too timid to make a move. I was just an insecure, suburban teenager from Helsinki in Finland. I did well in school and had no real worries. But I was desperately searching for something to kick-start my life. I had hoped to find a sport in which I might shine, but I was no good at athletics, basketball, football, or ice hockey. Quite simply, I was a mediocrity. Something had to change! As I pedalled home, I convinced myself, it's now or never.

A few days later, I saw my chance. The sports society of our school, Ågeli, needed a new chairperson. I applied and was elected. Many students in our school were good at sports, but they had not been active in school competitions. This was going to change. We took part in tournaments and started to have some success. I remember especially our home match against Grankulla, a fancy school. It was winter, and these were skiing competitions. The Grankulla captain was dressed in an elegant fur coat. He looked at me and sneered condescendingly. I cannot remember how I was dressed myself, but certainly not in furs. His glance said everything: "These guys will be a piece of cake." Nope, Ågeli won, and I felt good.

When the position of chairperson of the high school student society became vacant, I was ready. Who volunteers for these tasks anyway? I applied, and again, I was elected. This time I had some experience. I knew that in order to have some clout, the society needed an effective board. Ågeli's pupils came from a large area, and most of them travelled to school by train. Many were from working-class families, including some tough guys. They used hair pomade and wore flat hats and winklepicker shoes. In our Finnish-influenced slang, we called them *lättähattuja* (flat hats). At weekends, these guys went to dance at Helsinki Fair Hall. I knew that I had to recruit someone from the *lättähattu* gang to the board, and, in the end, I succeeded.

We also revived the school magazine, Noctiluca, which became our mouthpiece. Ralf Nordgren, who later became a writer, was chief editor. We sent the magazine to other schools to inform them about our activities. Our school dances, open to all Helsinki high school students, became a highlight. Ågeli had its own school band with Henrik Nyman and Seppo Finne, who both went on to be well-known jazz musicians. Also, our bass player, Jacke Söderman, later became a celebrity as Finnish Minister of Justice and European Ombudsman. The band was the main attraction of our dances, but I'm sure the dim dance floor lighting also played a role.

Ågeli was a Swedish-language school, and I belonged to the minority of Swedish-speaking Finns in Finland. We had had little contact with the two Finnish-language schools in our neighbourhood, but now we organised joint parties with them. All these events helped to boost unity and togetherness in our school.

The change became manifest when we organised a movie screening for our student society on an ordinary weekend evening. Although drugs were not a major problem in Helsinki at this time, we wanted to show a French feature film about drug addicts. There would be a discussion after the movie. We were not expecting a large audience, but the outcome was sensational. We managed to attract almost all the students at our school to a cultural event! Although most students had taken the train home after the school day, they came back for the evening. I could not believe my eyes; the hall was full. For me, there was magic in the air.

This was a formative experience for me. I realised that this success had to do with belonging and togetherness. Ågeli had become special. This magical feeling would guide me throughout my life.

***

My father, Lennart, was a physicist. A farmer's son from Vörå in the Swedish-speaking part of the province Ostrobothnia on the west coast of Finland, he was the first of the family to have an academic career, finally becoming a professor of physics at the University of Helsinki. My family had been farmers for at least 400 years – as far back as the family could be traced. When I was one year old, we lived in Denmark; my father worked at Niels Bohr's institute in Copenhagen as a visiting researcher. But he interrupted his stay there when the Soviet Union attacked Finland in November 1939; he went home to serve his country.

After the war, Lennart and our whole family spent a sabbatical year at Princeton in the renowned Institute for Advanced Study led by Robert Oppenheimer. Among the researchers working at the Institute was Albert Einstein. My younger brother Tom and I (an eleven-year-old at the time) knew that Einstein walked to work every morning. Once, we positioned ourselves expectantly in front of the entrance and stopped him when he arrived. Einstein exchanged a few friendly words with us and allowed us to take a photograph, which still hangs on my wall at home in Dresden.

In Helsinki, many famous physicists came to visit us. My mother, Rut, liked cooking and hosting dinners for the guests. She claimed that, unlike her, the wives of other physics professors shunned this task. Lucky for me, because my dream was to become a physicist myself.

During my last year in school, I had a long discussion with my father. "Do you still want to become a physicist?" he asked. When I said yes, Lennart looked at me and finally said "Kai, I don't think you have it in you to become a physicist." It took a while for this to sink in, but strangely enough, what I felt was mainly relief. Deep inside me, I'd had my doubts but suppressed them. Certainly, I was no genius at mathematics or physics. I could not see the solutions to complex mathematical and physical problems in a flash. Being best in class in Ågeli was not enough.

3

Lennart continued with a proposal, "Why not study medicine? Then you could become a researcher if you like, and if research doesn't appeal to you, you'd still have an interesting profession." With these words, he lifted the burden off my shoulders. Relieved, I followed my father's advice, and I'm eternally grateful to him for seeing both my weaknesses and my strengths. In his pragmatic way, he gave me the kick I needed. He was a farmer's son, after all.

I started medical studies in Helsinki while continuing to live at home. My brother, Tom, was studying architecture, and we lived together in the basement of our family's house. There were two rooms with separate entrances; it was an ideal arrangement. Tom's interests were completely different from mine; he and his friends kindled my interest in art, architecture, and music. Also, we continued to enjoy our mother's cooking and profit from our dinner-table discussions, which covered every topic and were loud and lively. Our sister, Majlen, who was eight years my junior, was impressed by her elder brothers, of course. At least, we thought so. But she was clearly irritated by our incessant talking and complained that we gave her no chance to get a word in.

In parallel with pre-medical studies, almost immediately, I started doing research, spending two summers at Karolinska Institutet in Stockholm, Sweden. First, I worked with Staffan Magnusson, who was investigating blood coagulation. The department head was Erik Jorpes, a world-famous biochemist who had introduced the use of heparin as a drug against blood clotting. Jorpes was a colourful Finn born on the island of Kökar. He studied medicine in Helsinki, participated in the Finnish Civil War during the country's transition from part of the Russian Empire to an independent state in 1918 on the side of the 'Reds' (the Finnish Socialist Workers' Republic), and ended up in Soviet Russia after the victory of the 'Whites'. He quickly realised the catastrophic mistake he had made. Communism was not what he had imagined, and he managed to escape to Stockholm. Magnusson married Jorpes' daughter and became a professor in Aarhus, Denmark. We remained friends and often met later in life. During my second summer, I worked with Bengt Samuelsson, who studied lipids and prostaglandins. I ran his complex apparatus for distillation and purification, which covered an entire wall with glass flasks and tubes. Luckily, I did not break any of the intricate laboratory glass contraptions. Samuelsson later received the Nobel Prize in Medicine and became the rector of the Karolinska Institutet. These two summers were an inspiring experience, nurturing my interest in science.

***

Together with my fellow medical student, Lars Runeberg, we did a special assignment for Ralph Gräsbeck, who was an assistant teacher at the Department of Medical Chemistry at the University of Helsinki. Gräsbeck was a clinical chemist studying vitamin B12 absorption, and the task he gave us was to surgically remove the stomachs of rats and measure B12 uptake and transport into the blood from the intestine. I managed to perform the gastrectomies, but my surgical skills were poor. I realised that I should not become a surgeon – one more profession I could strike off my list! Still, our study was published in The Lancet, which gave Lars and me a feather in our caps. I started dreaming that maybe Runeberg and I might do research together in earnest. A troika would be the ideal group, I imagined. In case of a conflict between two, the third partner could always mediate. I was already thinking of a third candidate, Amos Pasternak, who was a fellow medical student interested in research. Neither of them had any idea of my plans for them.

The Medical Chemistry department was rather special. It was led by Paavo Simola, who had been a prominent biochemist in the 1930s. He lost his daughter in a traffic accident, however, and this tragedy brought him down. Despite his personal problems, including alcoholism, Simola was able to continue as department chair until

1961. Although he had lost the joy of research, he had retained his sound judgement in selecting assistants. He was capable of attracting talent to his department.

At this time, biochemical research was not particularly expensive, and the resources of the institution were sufficient to support the research. Simola gave us free rein. Together, the young medical students who were attracted to the department created a stimulating atmosphere, although the laboratories had hardly any daylight at all. Many of the researchers working there would later become well-known. Among them were Kari Kivirikko, who built up an international research environment in Oulu; Aarne Raina, who became a professor of biochemistry in Kuopio, and Reijo Vihko, who became a professor of clinical chemistry in Oulu and later was President of the Academy of Finland. There was something enigmatic about the atmosphere of the department, but it worked. Simola suggested research topics for his teaching assistants, which some of them followed successfully throughout their careers. Beyond that, everyone decided for themselves. Simola did not interfere, and my impression at that time was that there was no leadership at all. Interactions between the researchers were positive, however. Most importantly, everyone was young, and this certainly helped. With hindsight, I think that Simola probably played a greater role than I had been able to see as a young student.

I was examined in medical chemistry by Simola shortly before his retirement. The exam was held in his office. The curtains were drawn, and the atmosphere was dark, as it was in the rest of the institute. A violin was lying on a table, and I wondered if Simola played when he needed to cheer himself up. He had to have some secret. In a mysterious way, he had managed to create a research environment that was more creative than many I encountered later.

After obtaining my Bachelor's degree, I embarked on my doctoral dissertation in parallel with clinical studies. I chose to continue with

Gräsbeck as my thesis advisor. Vitamin B12 was a long-established focus of medical research in Finland because, at that time, most people in eastern Finland were infested by the broad tapeworm, an intestinal parasite that causes B12 deficiency. The habit of eating raw fish contributed to the high prevalence of this disease. A tapeworm can grow up to 15 meters long, and one person might be infected with several worms simultaneously. The record was 330 meters of tapeworm in one patient.

Tapeworm carriers were permanently hungry. When the endemic infestation was at its worst, it is said that the worms sucked 60,000 kilograms of butter daily out of the Finns. The worm carriers suffered from headaches, fatigue, vertigo, stomach problems, and, of course, weight loss. Opera singer Maria Callas is said to have taken a tapeworm treatment to lose weight. She was 108 kilograms when she started her cure, and with the help of the worm, she lost 40 kilograms! More seriously, the B12 deficiency that results from the tapeworm consuming vitamin B12 from the food can lead to pernicious anaemia, a disease that is lethal unless treated. It is no wonder that B12 research had become such a big field in Finland.

Bertel von Bonsdorff was the doyen of the Swedish-speaking medical community in Finland. He had worked with Willian Castle at Harvard, who had postulated that the cause of pernicious anaemia is that the gastric juice lacks a so-called intrinsic factor necessary for B12 uptake into the bloodstream. Now, Ralph Gräsbeck wanted me to isolate and purify this factor from gastric juice. His earlier experiments suggested that the factor was a protein. He had calculated that we needed forty litres of gastric juice to isolate enough of the factor to characterise the protein. That is a lot of gastric juice! What a challenge. How could we manage that?

Gräsbeck had moved to a new laboratory in the Minerva Research Institute founded by Bertel von Bonsdorff, Bror-Axel Lamberg, and himself. The laboratory was housed in the small Methodist Konkordia

Hospital in Helsinki. Von Bonsdorff assembled a circle of accomplished medical researchers who also worked as docents (adjunct professors) at the University and were dedicated teachers.

The physicians gathered regularly at the meetings of the Finnish Medical Society, *Finska Läkaresällskapet,* in the House of the Estates. The meetings were open to young students like me, and I attended with great pleasure. The discussions after the lectures were lively; the curiosity of the participants was especially striking. The members of the society were inquisitive and wanted to be informed about progress in medicine. We young doctoral students were also encouraged to give lectures as soon as we had some data to present. The atmosphere was welcoming and inclusive, and we felt that we were members of a community. Von Bonsdorff was always enthroned in the front row. He exuded dignity and evoked respect but rarely asked questions or made any comments. I would have liked to discuss my work with him or to hear about his experiences doing research in Boston, but I never dared approach him.

My father, who had grown up in simple circumstances on an Ostrobothnian farm, had difficulties coping with the Swedish-speaking academic nobility in Helsinki. Its discreet elitism made him feel inferior. What amazed me was that he never had the same feeling elsewhere in the world.

***

The Finnish-speaking medical elite in Finland was dominated by a group that called itself the Bear Gang, *Karhukopla*, led by Esko Nikkilä. Like von Bonsdorff, Nikkilä was a professor of internal medicine and head of the clinic at the University Hospital in Helsinki. Their agenda was to improve the quality of healthcare and biomedical research in Finland. They reformed the medical journal Duodecim, making it a medium for the continuous education of the medical profession. They ensured that the articles were scientifically well-founded and equalled the highest international standards. They felt strongly about medical research and also tried to raise the level of doctoral dissertations. Like a collective Cerberus, the gang watched over the Medical Faculty in Helsinki. If a thesis submitted to the Faculty Council meeting were judged not to be of the required level, it would be rejected. They also wanted to ensure the optimal use of the limited budget for healthcare. The Bear Gang had great ambitions.

The Bear Gang arranged seminars at the Department of Zoology, where one of its members, the zoologist Sulo Toivonen, was working. To broaden the scope of the gang, they recruited this prominent representative of basic biological research. Toivonen and Lauri Saxén, another member of the Bear Gang, collaborated in their research, so questions that had no direct medical application were also discussed. Both were developmental biologists working on the morphogenesis of the kidney. They also had contact with the renowned cancer researcher George Klein at Karolinska Institutet, which saw a continuous stream of international biomedical celebrities coming to Stockholm. Scientists who were invited to lecture at Karolinska seldom declined because this was where the Nobel laureates were selected. Sometimes these lecturers agreed to take a detour via Helsinki to give a talk at the Department of Zoology. In this way, the seminar participants, including young scientists like me, got up-to-date information about research from around the world. These seminars were high points on my calendar. I felt like a member of an elite club, and it was easy to talk to everyone. In retrospect, it was also amazing that clinical professors who were heads of clinics were so accessible. For me, it was encouraging that the other doctoral students who attended the seminars also wanted to become researchers. It strengthened my conviction that I had made the right choice in combining my medical studies with research training.

***

When I began my studies, two main fields attracted talented students in Finland: the Faculty of Medicine at the University of Helsinki and the Department of Technical Physics at Helsinki University of Technology. Many applied, but few were accepted. Antti Vaheri, one of my fellow students who also became a researcher, once told me that he had hesitated between these two. Since these areas are so totally different, it could only have been the elite aura of these institutions that explained the attraction. The consequences for Finland of this channelling of young talent into medicine and technical physics were profound.

The success story of Nokia rested on graduates from the Department of Technical Physics. In the 1950s, Professor Laurila, who was the promotor of these studies at Helsinki University of Technology, had already initiated a project that aimed to build the first computer in Finland. This was accomplished in 1961. The computer, named Esko, was not a commercial success, but the project laid the foundation for the subsequent development of computer and information technology in Finland. Another key figure was Björn Westerlund, one of the big names in post-war Finnish industry. As Chief Executive Officer of the Finnish Cable Factory in Salmisaari, a small neighbourhood of Helsinki, he founded an electronics division that hired young physicists. The Cable Factory had a solid revenue from exports to the Soviet Union, which enabled the electronics division to recruit brilliant technical physicists and to support a think tank, sowing the seeds of Nokia's later success. In 1967, the Cable Factory merged with Nokia, which until then had been mainly a rubber and wood processing company. In 1982, Nokia's first mobile telephone was released. It weighed nearly 10 kilograms.

When Westerlund founded the electronics division, he hired the prominent mathematician Olli Lehto, who later became Rector and Chancellor of the University of Helsinki. In jest, Lehto called the Cable Factory 'the University of Salmisaari' as so many of the leading names in Finnish information technology started their careers there. Once again, it was the conglomeration of young researchers and engineers that promoted their collective success. In retrospect, it appears incredible that a private business in Finland could afford to build such a research team and give them the freedom to perform groundbreaking research and development for such a long time. None of this would be possible today. But the investment paid off and created a commercial revolution in communication technology.

***

The Faculty of Medicine attracted talented young people who wanted to become both physicians and research scientists. It educated professionals for its own needs, for regional requirements, and for the whole country. In addition to Helsinki and Turku, medical faculties were founded in Oulu, Kuopio, and Tampere. Many academics in Helsinki criticised this expansion and feared that there would not be enough young talent to fill all the positions needed and that the overall quality would drop. In the long run, it turned out that this investment in the medical education system was wise. It laid the foundation for reform of the entire healthcare system in the 1960s and 1970s. Finland learned from the mistakes of Sweden by introducing a hospital hierarchy in which only the university hospitals were given the resources to provide comprehensive medical services with all the necessary equipment. Five university hospitals were located throughout the country and could therefore provide modern health care to the entire population. Regional hospitals had to focus on normal patient care. The most demanding patients were left to the university hospitals to diagnose and treat. This reform led to significant savings when compared with Sweden, where the regional hospitals could acquire expensive equipment even if they lacked competent staff and enough patients to examine. Medical practice in Finland was on a high level because it was supported by a strong research base, initially in Helsinki but later at the other medical faculties as well. The establishment of healthcare centres across the country was a third pillar of this reform, which contributed to the efficiency of Finnish healthcare.

One key to the success of the Finnish healthcare reforms was that they were planned by highly educated experts who were well-connected to leading institutions in Europe and the USA. Many medical professors had done research abroad and established important contacts. Kari Puro, who led the great healthcare reform in the 1970s as undersecretary at the Ministry of Social Affairs and Health, came from Paavo Simola's talent forge at the Department of Medical Chemistry in Helsinki.

The results of the reforms were dramatic. Finland had long had a much lower life expectancy than the other Nordic countries. Today, Finland rides high in the international ranks. About 9.5% of gross domestic product (GDP) is spent on healthcare, and the average life expectancy is eighty-five years for women and eighty for men. Sweden has a similar life expectancy, but it spends 10.9% of its GDP on healthcare. The 2% difference corresponds to over four billion euros per year saved in Finland. Compare that with the USA, where the life expectancy is only eighty years for women and seventy-five for men. To achieve these miserable statistics, the USA spends more than 18% of its GDP – almost double what Finland pays out!

**

In 1963, I started working in earnest on my thesis, supervised by Ralph Gräsbeck. My task was to characterise the intrinsic factor needed for vitamin B12 uptake, which, according to Gräsbeck's calculations, would require me to collect forty litres of gastric juice. A tube had to be inserted through the oesophagus into the stomach, and then some ten millilitres of gastric juice could be sucked up from each subject. The principle was simple enough, but how would it be possible to find enough subjects willing to donate their gastric juice for our project?

Gräsbeck hired a nurse to tour the prisons in Helsinki in search of volunteers. Whenever the nurse came knocking, there was a long line of prisoners waiting for her. Today, this way of collecting samples would be considered unethical. Times have changed.

I expanded my thesis project to include analyses of other body fluids as well. To collect enough saliva, I asked the professor of obstetrics if I could have my fellow students collect their own saliva during his lecture. Our professors often took a positive attitude towards students who were doing research during their studies. Many clinical professors had, themselves, worked on their dissertations while studying. It was a hilarious sight: sixty medical students chewing paraffin to stimulate secretion and spitting into beakers on ice to cool the saliva, all while trying to concentrate on the lecture. At the end of the lecture, I pooled all the saliva collected and analysed the B12-binding proteins in the lab.

One summer, I worked in Vasa and stayed with my grandmother. She lived in a beautiful old wooden building in the centre of the city. My grandmother was gentle and friendly, and it felt calming to be with her. The stairs up to her apartment were covered with linoleum, which had a distinct smell. Whenever I sensed the same smell later, I always thought of Grandma. It reminds me of Proust's novel, 'Remembrance of Things Past', in which he describes the power of the sense of smell in evoking strong memories. During the war, Grandma kept a pig in the basement of the building where she lived. This was forbidden. One day the pig escaped and started to explore Vasa. Grandma followed anxiously, trying to persuade her pig to return home. But it was useless; the pig ended its city tour at the steps of the police station, and there it remained until discovered. Poor Grandma; all that work for nothing.

In Vasa Central Hospital, I worked together with the head of clinical chemistry, Wolmar Nyberg, one of von Bonsdorff's former B12 students. My work with rats was over. I developed a method for measuring the uptake of B12 from the human gut into the bloodstream. My method entailed inserting a plastic tube ending in a little steel ball into the intestines of the patients. The tube had a balloon that could be inflated with air to occlude a section of the intestine, into which I could then inject vitamin B12 and a preparation containing the intrinsic factor to measure its potency. It was not easy to position the tube at the right point in the intestine, so I had to X-ray the patients to check that it was correctly located. The patients were so grateful that I spent so much time with them, although I had explained that I was not treating them but performing experiments that other patients might one day profit from. I was touched by their enthusiastic participation even though they derived no benefit from the procedure, but, at the same time, I was troubled by having to do research on patients. After my brief time at Vasa Central Hospital, I decided that basic research should be my future profession.

***

When the autumn term started, the chairmanship of the Swedish-speaking medical students' club, Thorax, became vacant. I volunteered and was elected, just like I did at school. My first task was difficult. By tradition, Thorax produced a musical show each year. The show was usually performed not only at the annual gala of the club itself but also for a wider audience in Helsinki and often even in Stockholm and Gothenburg. The Thorax shows were ambitious, with opera and operetta arias and other songs woven into a narrative. Now it was my duty to ensure that a new show would be performed during my term in office. The problem was that the group that had written and staged the shows for years had decided to retire. They were finishing their studies. Their leading figure had been Claes Andersson, an all-round talent who later became a well-known psychiatrist, poet and writer, jazz musician, and politician. I would have to conjure up a new Claes Andersson, which was, of course, impossible. Neither were we allowed to seek help from other faculties – the Thorax shows had to be produced by medical students.

Luckily, I happened to hear about a promising candidate: a student at the Sibelius Music Academy who wanted to swap to medicine. He had applied to the faculty but had not passed the entrance tests. I persuaded a fellow medical student to prepare him for the next selection. If successful, I promised him that he would be rewarded with a stipend that Thorax awarded each year. The plot was successful. The Sibelius Academy student became our new show master and the musical leader of a long line of fine shows, which delighted us and our friends for years.

We initiated a dialogue with Thorax's sibling society for Finnish-speaking students, *Lääketieteenkandidaatiseura* (LKS). Until then, the two societies had led separate lives. A fellow student, Juhani Hyvärinen, was the chairman of LKS, and we started thinking about what our two societies might do together to activate our medical students. We devised an action along the lines of the *Teekkaritempaus*, a campaign by engineering students in Helsinki that always had a social purpose. We started a campaign to fight the tapeworm in eastern Finland.

The Finnish pharmaceutical company Medica had launched a new drug that eradicated the tapeworm from patients. If we could treat enough patients to eliminate humans as hosts in the infection chain, the parasite would disappear. The tapeworm's life cycle requires that its eggs pass with human faeces into lakes and rivers, where they are ingested by fish, which are then eaten by humans. The population in eastern Finland was so used to the worm that they regarded all the unpleasant symptoms of the infestation as something entirely normal. Now, finally, they could see how much better their quality of life would be without worms. Thorax and LKS together decided to go for the cure, and this was indeed the beginning of the eradication of the tapeworm in Finland. We organised parties of medical students that went around selected areas in eastern Finland, distributing the drug to people. The campaign was successful. After some years, the worm was gone.

As chairman of Thorax, I had the privilege of attending the 1962 Lucia Ball of the Society of Medical Students at Karolinska Institutet in Stockholm. According to tradition, that year's Nobel laureates in Physiology or Medicine, Francis Crick and James Watson, and in Chemistry, Max Perutz and John Kendrew, were invited to the ball. All four were pioneers of molecular biology. I was expected to give a speech. I was petrified, knowing how elegant and humorous the Swedish students were when they gave their speeches. For me, talking publicly was an effort. I simply lacked the fluency and the touch that was needed. This time, I had read an article about 'organismic biology' in the Journal of Theoretical Biology and managed to weave it into my presentation about the biology of the future. Seemingly by accident, I mispronounced it 'orgasmic biology' and the audience burst into laughter. My evening was saved.

***

Finally, we managed to collect forty litres of gastric juice and embarked on the task of isolating the important protein – the intrinsic factor. We had planned and practiced how to do the purification, and we had a protocol to follow. Some steps of the isolation process stretched over twenty-four hours. I slept in the lab with an alarm clock as my only companion to make sure that nothing went wrong. Our material was too valuable to allow any technical mishaps.

The work bore fruit. We successfully isolated our intrinsic factor, the B12-binding protein, and could characterise its chemical and physical properties. Since my other investigations were completed about the same time, I wrote up a ninety-three-page monograph about my B12 work. I defended my doctoral thesis in 1964, the same year I finished my medical studies. I was going to a B12 symposium in Hamburg together with Bertel von Bonsdorff, Ralph Gräsbeck, and

Wolmar Nyberg. We were all about to check in at Helsinki Airport when I realised "Oh my god, I've forgotten my passport!" Very embarrassed, I had to return home to fetch it. I can still recall von Bonsdorff's expression when he turned around and looked down at me. But I made it, in the end, to Hamburg.

To afford the trip to Hamburg, I had applied for a travel grant from my student fraternity, Nylands Nation. The discussion about my application that unfolded at one fraternity meeting was hilarious. The fraternity chief held that it was preposterous to grant travel funds for a symposium. Why? Well, because 'symposium', according to him, was just the Greek word for a drinking party. He was referring to a famous painting called 'Symposium' by Axel Gallen-Kallela from 1894, which represents the artist together with the composers Jean Sibelius and Oskari Merikanto and the conductor Robert Kajanus in a state of heavy intoxication in Restaurant Kämp. The objections of the chief notwithstanding, I was granted the funds. At the symposium, I had the opportunity to meet the *grande dame* of the field, Dorothy Hodgkin, who, a few years earlier, had determined the structure of vitamin B12 by X-ray crystallography. She was impressive and unassuming in equal degrees. Margaret Thatcher had done research in her lab in Oxford as a chemistry student in the 1940s, apparently without being much influenced by the personal qualities of her mentor.

Now and then, I attended the events of Nylands Nation to meet students from other fields, but I was also looking for a partner. There was no online dating service then, so I had to find candidates in person. I had my antennae out but had not had much luck so far. When my studies were nearing their end, I set my eyes on Carola Smeds, who was the hostess of the fraternity that year.

She was always surrounded by other students, and it was not easy to approach her. I persisted as I began to realise that this was developing into something serious, at least for me. One day, I was lucky enough to bump into Carola in town. I managed to ask her if she would like to accompany me to the zoo. To my delight, she answered yes. I was head over heels in love and bewitched by Carola. For me, it was a day without comparison. Magical things can happen, indeed. Carola kept me at a distance for a while, but I was undeterred and, finally, we became a couple. It was fantastic!

***

Throughout my life, I have taken an interest in societal issues. I had read C. P. Snow's book 'The Two Cultures and the Scientific Revolution' about the two cultures, 'science' and 'the arts', and found the book so interesting that I decided to write an article in the main Swedish-language newspaper in Finland, *Hufvudstadsbladet*. I wrote about the neglect of scientific culture and gave some thoughts on how it might be revived in Finland. To my delight, the article was published on 17 June 1965, a few days before midsummer when Carola and I were to be married.

One week earlier, my friends had organised a stag party for me on an island far out in the archipelago east of Helsinki, where the Simons family rented a summer house. As befits such a party, the vodka was flowing. I danced naked outside the sauna on the cliff in a performance of the dying swan. There is somewhat blurred evidence of this event in the form of a large, black-and-white photograph on the wall at home. From there, the whole gang was taken to the mainland in a big fishing boat, but I have no idea how we ended up there. I briefly regained consciousness when I was taken into police custody, where I was placed in a cell for the rest of the night. In the morning, I woke up to quiet singing. I was not alone. My two cellmates looked like regular customers. One of them was singing melancholy songs about his life, warming my heart. Strangely enough, it was a pleasant feeling to wake up in a prison cell. When I was being released, I noticed that my watch had disappeared. The police had requested me to take it off when I was handed in. Now they checked who had arrived at the same time and managed to catch the thief before he had time to sell the watch.

The thief had had such presence of mind that he had snatched my watch off the desk without the police even noticing. Of course, I did recover in time for the wedding with my beautiful, beloved Carola, in the company of our family and friends.

There was just one small fly in the ointment. On the morning of our wedding day, a rejoinder to my article about the neglect of scientific culture, written by the artist Carl-Gustav Lilius, was published under the headline 'The Delusions of Science'. Unfortunately, I read it. It began "A fanatic, new-born dogmatic belief and an astonishing blindness to the delusions of science characterise Simons' article." Lilius hit with a sledgehammer and went on to demolish my ideas with the same force from beginning to end. The invectives kept me company while I was standing at the altar waiting for my bride.

My first major article outside my own field had elicited such bitter criticism. Lilius managed to start a heated debate about science contra everything. Much later, in 2006, Lilius' widow, Irmelin Sandman-Lilius, wrote a biography about her husband, which included "Kai Simons' text of 18 July", my next instalment in the debate, "is interesting and balanced, he develops his ideas about the dissemination of knowledge and gives a fascinating picture of the revelations of modern physics: if I saw the floor as it is 'in reality' – a void with atomic nuclei flying here and there in complex patterns surrounded by clouds of electrons – I would hardly dare to stand on it. Increasing numbers of various experts rushed to join the debate. The titles of their contributions give the impression of fluttering banners in a protest march … … Everyone did not march in step."

As we sailed off for our honeymoon in Denmark, standing on the ship's deck in the wind, we were in a completely different world. The debate was just a faded memory. We rejoiced in our common future. The ferocity of this debate, however, was excellent preparation for my later life, when I was hit by a vicious attack on my novel idea of cell membrane organisation, the lipid raft concept, that went on for years.

***

Now I faced a question: what should I do as a newly graduated Doctor of Medicine and Surgery? I wanted to go to the USA, the promised land of biomedical research. Almost all European graduate students who were planning a career in biomedical research went there for postdoctoral training. Finland had paid its debt from World War I to the USA, and this was presently being repaid as research scholarships. No less than six such National Institutes of Health (NIH) scholarships were at the disposal of the biomedical sector in Finland every year. Hilary Koprowski, who was director of the Wistar Institute in Philadelphia, came to Helsinki every year for talent scouting. I met him at an event in the home of Tapani Vainio, one of the Bear gang. Koprowski was a stimulating scientist, and many Finns had decided to go to the Wistar, but I wanted to experience another environment.

My work on vitamin B12 had piqued my interest in cell membranes. I found it fascinating how the Bl2 molecule must bind to a protein – intrinsic factor – which then binds to the plasma membrane of intestinal cells in order to pass across the cell. The more I read about membranes, the more enthusiastic I became. Not only was the cell enveloped by a membrane but all the compartments within the cell were bound by membranes, too. This was exciting. Also, I realised how little we understood at that time about cell membranes, which made me even more interested in pursuing research on them. Now, I wanted to identify a laboratory where I could work on cell membranes.

I was searching for simpler experimental models than the intestines of humans and animals, but I had no one to whom I could turn for advice. Finland was not centre stage in the molecular life sciences. I searched the scientific literature, but this field was new; I could not find a suitable lab. Most research on membranes focused on mitochondria, the power plants of the cell, whose membranes are essential for energy production. I was looking for less well-trodden paths. Finally, I settled on Rockefeller University in New York, where Alexander (Alick) Bearn was working on a problem I believed was close to my own interests.

Bearn accepted me, and I managed to get an NIH research scholarship. In August 1965, Carola and I travelled to New York. We took the ferry to Stockholm, the train to Gothenburg, and the passenger ship Gripsholm across the Atlantic. On the boat, we met several Swedish emigrants who had been on their first return visit to the old country. It was interesting to hear how disappointed they were. They never imagined that Sweden would develop into a country with such a high standard of living. So much was better at home in old Sweden than in the USA! But this did not discourage me. Finland still lagged so far behind in science that I was certain I would return home as a fully-fledged scientist.

Chapter 2: New York

For the second time in my life, I was welcomed by the Goddess of Liberty – the first time had been as a child, when my father took the whole family to Princeton. Then, we had travelled on a cargo ship and experienced such a terrible storm in the middle of the Atlantic that I had sworn never to cross by boat again. The night before the storm, I had eaten melon for dessert, and it would be twenty years before I ate my next melon. Gripsholm was a much bigger vessel, and this time we had no problem with seasickness. For both Carola and me, it was a pure delight to cross the Atlantic on a passenger ship. The only misfortune occurred when we arrived. We were looking out over the docks at Manhattan and saw how the luggage was carelessly thrown on a slide going down to the wharf. To our dismay, we recognised our own bags, which had burst open, causing clothes to tumble out over the slide. Our first impression of Manhattan was that of hectic, bustling life – totally different from what we were used to in Helsinki. The taxi drove so madly that we were clinging to the seat.

Rockefeller University is located in Manhattan between East River Drive and York Avenue. We were lucky to get an apartment on campus in Sophie Frick Hall. It consisted of two rooms and a bathroom. There was no kitchen. Carola, who has a practical mind, decided to buy a countertop burner. We kept that in the bedroom and washed the dishes in the bathtub. Since cooking was prohibited, we had to air the room properly and make sure not to be caught.

The research group where I worked was international. My boss, Alick Bearn, was a jovial Englishman. But, unfortunately, cell membranes and transport, which I had hoped to work on, had been taken off the table. The team's research was focused on human genetics, especially on the genetic variation of blood proteins. Although this was a great disappointment, I had no alternative but to choose a project that would fit into the programme of the group. I decided to isolate and characterise the proteins on which Bearn's group was working. This meant that I would continue with the same type of work as in Helsinki. I decided to deepen my skills as a protein chemist and learn new methods. At Rockefeller, I became an expert in electrophoresis techniques for separating proteins by using electric fields. I hoped that specialising in protein chemistry would help me to get a position back in Helsinki. My initial disappointment was assuaged by being allowed to buy a novel preparatory polyacrylamide gel electrophoresis machine developed for isolating proteins. This contraption became the apple of my eye – it was hard to handle, but I learned how. This made me feel good because it was a machine that almost no one else had. Purification of proteins is an important part of biochemistry, so I soon convinced myself that this was worth the effort.

I wanted to participate in an advanced course in physical protein chemistry organised by Gerald Edelman, later a Nobel laureate. I contacted him to register for the course, but it was not so simple. Edelman wanted to know my background; when I told him that I had studied medicine, he would not have me. I did not give up and added that I had taken biochemistry and mathematics alongside my medical studies at the University of Helsinki. I finally asked him to give me a week to delve into the subject, and then he could check my knowledge. I got my week and was accepted onto the course.

Our laboratory was in the same building as the hospital of Rockefeller University. It was a research hospital, founded in 1910, which had played an important part in the development of clinical research in the USA. Donald van Slyke, one of the founders of clinical chemistry in the USA, and Henry Kunkel, a pioneer of clinical immunology, had both worked there. Alick Bearn asked me to take the exam that was required to be allowed to work as a doctor in the USA. He wanted me to be on call at the hospital during nights and holidays, which would give me extra income and contact with the clinical research going on there. The general theme of the research was obesity. The researchers wanted to understand what happens when people put on weight and when they lose weight. One important finding was that when a person loses weight, the adipose tissue decreases in volume, of course, but the number of fat cells does not – the cells just shrink. This triggers a horrendous cacophony of signals from the fat cells, which are screaming for more fat. Therefore, the patients were constantly on the lookout for food and often asked their visitors to bring with them their favourite snacks. As a result, the hospital had to search all visitors to prevent them from smuggling food to the patients on diets. This was the situation in the 1960s. Since then, obesity has become a global scourge that we still have not learnt to control. I had no clue then that prevention of obesity was going to be the subject of the final phase of my career.

Rockefeller is not a normal university. It offers no undergraduate teaching and only accepts doctoral students, selected from those who have completed their bachelor's or master's studies at other universities in the USA and around the world. This made the atmosphere on the campus more ambitious and subdued than one might expect of an American university. The campus area had two dining halls. One was in Welch Hall, a building from 1929, famous for its architecture, and overlooking the East River. The food was served by waiters at set tables. The other dining room, in Abby Hall, was more modern, but had an equally dignified atmosphere. Having lunch with my colleagues dressed in a suit made me feel like a member of some important club. I suppose that was the idea. This chimed with the high quality of the research. Rockefeller has become a true innovation hub, producing twenty-six Nobel laureates.

***

When the opportunity to go to the USA came along, Carola was in the middle of her dentistry studies in Helsinki, which she hoped to continue in New York. To explore what was possible, she organised a meeting with the Dean of the Faculty of Odontology at New York University. Kindly but firmly, he made it clear to Carola that this was unthinkable. She was not eligible. All dentistry students were male. In Helsinki, most dentists were women – not so in the USA. Fortunately, Carola was soon accepted as a trainee with a dentist who had a practice on Park Avenue. Sometimes she assisted me with the purification of my proteins, and this was a great help because the methods were complicated.

Kåre Berg, the Norwegian in the lab, and his wife, Reidun, became our best friends. They, too, lived in Sophie Frick Hall. We often went out together, and the same rigmarole was repeated every time. After they had closed the door and taken a few steps, Kåre sent Reidun back to check that he had closed his desk drawer; every time it was the same procedure.

We also met another couple, Leevi and Kaarina Kääriäinen. Leevi was a visiting researcher at the Sloan Kettering Institute on the other side of York Street, opposite Rockefeller. It felt good to have Finnish friends in the metropolis, like a warm breeze from home. Leevi and Kaarina liked to play cards – bridge mostly. I have never been good at games. I like to talk and discuss, and I cannot focus on the game. But our friends would not accept this and insisted "Kai, please concentrate!"

Leevi studied viruses and held a position at the Department of Virology in Helsinki. When he told me about the Semliki Forest virus, the subject of his investigations, I became excited. By modifying a small piece of the host cell plasma membrane, this virus envelops itself in a lipid membrane as it exits the host cell. This virus membrane is much simpler than the complex membranes of the cell itself. Maybe we could collaborate to study it together as an experimental membrane model when we return to Helsinki? We began fantasising about this possibility, but how it could be realised was far from clear.

We had parties with lots of guests at home in our two small rooms. Once, we invited Lala Eriksson, a socialist acquaintance from Helsinki, who was working at Columbia Law School supported by a similar scholarship to mine. Alick Bearn and his wife, Margaret, came to the party in formal dress directly from another event. When Lala saw Bearn in a tuxedo, he got up on a table and gave an impassioned speech about the horrors of capitalism, imitating Alick's Oxford accent. But Alick had a good sense of humour. He started bickering with Lala and turned the whole incident into a joke. The next day, I waited for Alick's comments with dread, but he simply thanked me for a pleasant evening.

One day when Carola was walking to work on Park Avenue, she had a frightening experience. She heard a hissing sound and felt something nudge her back before there was a loud thump. She looked back and saw a bleeding person lying dead on the sidewalk. A man had committed suicide by jumping from the skyscraper just as Carola was passing. She was in shock and ran around in circles until she was taken care of by other pedestrians who had seen the whole tragedy. It was a narrow escape; if she had been two seconds slower, she would have been lying dead on the sidewalk herself.

To forget this unpleasant incident, we went to Mexico for a vacation. First, we went to Acapulco, which, at that time, was a haunt of the jet set. We had very little money but still dared to try a luxury hotel. Surprisingly, they had an inexpensive vacant room. We were blissful until we entered the room. All the hot exhaust air from the hotel's air conditioning blew towards it. The nights were terribly hot, but we endured and were rewarded with the best fish meal we have ever had: a big, grilled red snapper in a restaurant with a view over the Pacific Ocean. It is remarkable how such culinary experiences stay in one's memory.

On our way home from Mexico, we passed through Miami, where we rented a car and had an accident. From a side street, I turned into a main thoroughfare packed with cars. I was waved on, and I did not notice a low sports car that drove right into our car. After being examined at the nearest hospital, we were taken to City Hall, where I was taken to court and fined. We had not counted on fines. Our travel money was spent, and we had no reserves. Credit cards did not exist at that time. Even worse, I was not allowed to call my bank. The police officer who accompanied us gave us no choice; if we could not pay the fines, I would go to prison. When the policeman left for a moment, Carola and I saw our chance. We ran off to find a bank, take out money, and pay the fines. When we got back, the same policeman could not believe his eyes. Never before had any fugitives voluntarily returned to pay their fines. He had seen that we came from Finland and commented that this was something that only Finns would do. It was commonly known at that time that Finland was the only country that had paid all its war debts to the USA. It was more than lucky that we returned, however. While we were gone, the officer had already put out a search call for me. If we had returned to the hotel, the police would have been there to catch me. If I had been jailed after running away, I would undoubtedly have been expelled from the country, and my career at Rockefeller would have come to an abrupt end.

***

When Carola returned to Helsinki to continue her studies, I intensified my research. Now I had to produce results that could be published. Without publications, I would never get a position in Helsinki. Bearn's laboratory had lots of space, and, like most of those who worked there, I had a whole room at my disposal. Everyone was pursuing their own project, and there was no real coordination of the research. Other labs held group meetings every week, in which the members regularly reported what they had achieved. This was not the case in our lab, which led to problems for me. On the mail desk, I found an abstract describing one of my projects authored not by me but by Barbara Bowman, who was working in the adjacent room. I was shocked and went to my friend Kåre for advice. Angrily, Kåre suggested that I talk to Bearn and request that the abstract not be submitted. It was intended for a congress, and the submission deadline was the very same day. Barbara had already gone to Houston, where she would spend Christmas. I persuaded Bearn to contact her and inform her about the decision. That was easier said than done, but, at last, Bearn got the information that she was out shopping in a big department store. He managed to contact her through the speaker system of the store, calling her to the information desk. Barbara was terrified, thinking that one of her nearest and dearest had passed away, but it was only an abstract that had passed… into the garbage bin.

Relations between Barbara and me were strained for some time after this. She explained that she had started the project, but I never saw any results. We did our best to gloss over the incident. Each Friday after work, the whole lab went to the bar TGIF (Thank God It's Friday). After two Fridays, we had had enough drinks to be reconciled. The most important thing for me was that Barbara had lost interest in the project, so now I could keep it for myself.

Part of a scientist's everyday routine is to attend lectures and keep up to date with what others are doing. At Rockefeller, there were lectures every day, but some were more important than others. One important series was The Harvey Lectures, initiated in 1905. The lectures were given annually by renowned biomedical researchers, who regarded it as a great honour to be invited to talk to The Harvey Society. On one occasion, when I was in the audience, we were all waiting for the lecture to begin, but there was no speaker in sight. Suddenly, a group of gentlemen dressed in tuxedos entered, merry, noisy and tipsy. One of them introduced the speaker, a well-known molecular biologist called Norton Zinder, in a humorous and disrespectful way. Zinder then stepped up to the podium equally under the influence, but after a while, he collected himself and gave a brilliant lecture.

I continued my work on the project that Barbara had wanted to seize, and I developed into a competent protein chemist. This would be my main strength when I returned to Helsinki. Lab life in New York rolled on and became increasingly routine. We all had our own projects, so no real added value came from working together in the same lab. The only person who united us in any way was Minnie, our lab assistant. She went from room to room washing the glassware – the lab had no dishwasher. Minnie hummed and spread happiness around her wherever she went. One evening, Carola and I were invited to her home. She lived in one room cluttered with gadgets. Three TV sets were showing different programmes. They were probably on all the time, except when Minnie was sleeping, but maybe then too. We were shocked by how Minnie entertained herself, but this was only a foretaste of the coming digital revolution.

For me, the most important thing now was to scrape together enough data for an article that could be published in a scientific journal. Just like writers who write books for a living, scientists must write scientific articles that are read and cited by their contemporaries. After one year in New York, I submitted an article for publication. As is customary, I received reviews written by two referees selected by the editors of the journal. To respond to their criticisms, I had to do several additional experiments. The work took a few months, but finally, the article was accepted. With a paper in my pocket, I could apply for a position in Helsinki.

I was lucky; Finland had set up a programme to support young researchers who wished to return home after their postdoc. To be eligible for the programme, I needed to find a research institute willing to offer me laboratory space. The Faculty of Medicine had built a new complex at Haartman Street, which housed the Department of Sero-Bacteriology. I worked on proteins in the blood – naturally, an area of serology. This complex also housed the Department of Virology, which Leevi Kääriäinen would return to when he came back from New

York. Fantastic! This was indeed a beautiful fit. I contacted K. O. Renkonen, who was Professor of Bacteriology and Serology, and got my research position. Renkonen was a colleague of my father Lennart at the university, which probably did not decrease my chances.

What had I learnt from my experiences in New York? My greatest insight was that I did not want to do research in the way it was conducted at Rockefeller. Working alone on one's own project was unproductive and tedious. Biological problems are complex and must be addressed from various directions. This requires collaboration and coordination. Without leadership, this is impossible. Cooperation requires support. The strange thing was that the way we worked in Bearn's lab as individuals was more or less the rule at Rockefeller. To me, not only did this working style seem inefficient but, worse, it encouraged competition and elbowing within the group as well as between group leaders. Rockefeller produced twenty-six Nobel laureates, nonetheless. But I had the firm intention to work differently. I wanted to instil a spirit of community and cooperation in my own lab and in my surroundings now that I had been given the opportunity to lead my own research group at the Department of Sero-Bacteriology.

Chapter 3: Back in Helsinki

In the summer of 1967, Carola joined me in New York for the summer break before we took off with the low-cost Icelandic airline Loftleiðir to return to Helsinki. The plane landed at dawn, and when we got out of the taxi in Töölö, where Carola's parents lived, the feeling was almost magical. The light was so different from Manhattan. The air was clear, and the stillness enormous. Compared with Sophie Frick Hall, where we were exposed to continuous background noise from the traffic along East River Drive and York Avenue, Töölö was almost soundless. We had adapted to the noise and ceased to notice it until we came back to Helsinki. The silence was so intense that it almost hurt our ears. Good to be back!

That autumn, I began to work at the Haartman Institute, opposite the university hospital in the neighbourhood of Meilahti. The Department of Sero-Bacteriology was part of the new institute, led by K. O. (Pappa) Renkonen. His father had been *archiater*, the highest honorary title of a Finnish physician. Two of his sons worked in the same department. The elder son, Ossi, was already an acknowledged investigator in the field of lipid chemistry and had recently been appointed professor. When I came to the institute, Ossi was at Harvard working with Konrad Bloch, who had managed to synthesise cholesterol from scratch. Pappa Renkonen was an understanding and convivial head of department. He encouraged us to find our own way and supported us when necessary. His own research, he did alone. He wanted to know why more boys are born than girls. I don't know if he managed to solve that riddle.

Importantly, the department performed routine serological and bacteriological analyses for the university hospital and received compensation for its service. The profit did not go to the head of the institute, as in many other departments. Instead, Pappa Renkonen used the money to support the researchers working there. At that time, Finland was still a developing country with respect to research funding. Government support was among the lowest in Europe, comparable to that of Portugal. Thanks to Pappa Renkonen, I had the means to fund staff and running costs on top of my government grant from the Medical Research Commission. I was especially proud that he gave me money to buy an amino acid analyser, an expensive instrument by the standards of that time.

When founding a new research group, the toughest challenge is to attract interested and competent young researchers. I was fortunate when I recruited my first doctoral student, Carl Gahmberg. Carl had come to the institute for an interview with Thomas Tallberg, who had the room next to mine. Fortunately, Thomas was out when Calle arrived. Instead, he came to see me, and I had the chance to persuade him to join my group – successfully! I was also fortunate with my second recruit, Ari Helenius. He was a biochemist, not a medical student like Calle, and a perfect fit for what we were doing. He knew both physical chemistry and mathematics, which we physicians are usually less good at. Besides, he was dating my sister, Majlen. I had better behave and not make a fool of myself! Ari became both my brother-in-law and my long-standing collaborator. My third recruit was Henrik Garoff, a medical student. Calle, Ari, and Henrik all became internationally respected scientists. Calle became Head of the Department of Biochemistry in Helsinki and Henrik a professor at Karolinska Institutet in Stockholm. Ari became the successor to George Palade, the father of molecular cell biology, as Department Chair of Cell Biology at Yale. Later, he moved back to Europe to head the Biochemistry Institute at the ETH, the public research university in Zürich, Switzerland.

I was extremely proud to have been able to assemble such a super group. In addition, I hired a laboratory nurse, Hilkka Virta, as my assistant. Hilkka was quite a personality. She was as skilled as she was pedantic, but also hypersensitive. Besides, she was exceptionally demanding. Woe to him who did not clean up after his experiments! At one stage, it went so far that I told Hilkka that I could not live with her complaints any longer. Our roads had to part. But Hilkka begged to stay, and, of course, I gave in. We worked together until she retired.

To attract a powerful team, I had been forced to give a rosy picture of my projects. In reality, it was a bumpy ride in the beginning. We finished the experiments that I had started in New York. It was boring, but the studies had to be published. Publish or perish!

***

Kåre Berg had discovered a new genetic variant of a serum lipoprotein, Lp(a), which I introduced into our repertoire. Since I was Lp(a)-positive myself, I donated my blood, half a litre each time, from which to isolate the protein. My red blood cells were returned to me to prevent me becoming anaemic. I isolated the lipoproteins by using an ultracentrifuge, an expensive instrument that was available at the Department of Sero-Bacteriology. One morning when I came to the lab to take my samples out of the centrifuge, there was an intense smell of burning in the corridor. When I opened the door to the centrifuge room, the centrifuge was demolished. The new and expensive rotor, designed especially for the isolation of lipoproteins, had exploded! I was shocked. What had I done wrong?

I still managed to proceed with my experiments since the Department of Hygiene had a similar rotor. Without revealing my previous misfortune, I asked the department chair, Professor Rantasalo, if I could use the lipoprotein rotor of their ultracentrifuge. He gave his permission, and I used their machine for two weeks until there was a phone call from the Hygiene Department, "Simons, come here right away!" I could not believe my eyes; it was the same catastrophe once again: the rotor had exploded, and the ultracentrifuge was destroyed! I was desperate. What had I done!? This would be the end of my career. I called the company that supplied the instruments and asked what could have caused the explosions. They gave me no consolation – just the impression that I had not followed the instructions.

After two weeks of suffering, I received a happy message. Although the rotor was designed for the isolation of lipoproteins, it was damaged by the potassium bromide needed for the procedure. It was a construction error. I was saved! Since it took many centrifugations to damage the rotor so badly that it exploded, I was the first person in the world who had encountered this misfortune. My diligence had been my undoing. If I had been working in the USA, I would have suffered ignominy as an incompetent experimenter, apparently all fingers and thumbs. I was certainly not very skilful, but neither was I that incompetent.

Carola completed her dental studies and, in March 1970, our twins Katja and Mikael were born. This happy event completely changed our lives. It didn't occur to me to take paternity leave – that concept had not yet been invented, or, at least, I had never heard of it. Carola took care of the children; I didn't understand at all how much help she would need. Public day-care was not as well organised as it is today in Finland. My sister Majlen and Ari Helenius had their son, Jonne, about the same time, so half a year later, we decided to team up and hire a nanny. We managed to organise our lives together like this for almost ten years. At first, the children stayed mainly at our place, as our flat was larger. Later, we built a pair of bungalows close to my former school. When the bank saw our building plans, they refused to give us the loan we needed. The problem was that we had planned a sauna between the two houses with doors to both sides. We thought it would be nice if the children could move freely between the two homes. Only after long negotiations did the bank yield, maybe because my father Lennart knew one of the directors.

The connection between the bungalows proved very practical. The children could ride a tricycle between the homes, and if either family wanted to be in peace, it was easy to lock the door. Majlen and Ari had a second child, their daughter, Ira, and we were glad to have Gunvor from Sibbo taking care of all the children. She was a jewel.

By then, Leevi Kääriäinen, too, had returned from New York. His laboratory at the Department of Virology was one floor above mine. We planned to join forces to study the Semliki Forest virus. I was interested in the virus membrane, Leevi in the reproduction of the virus and how its RNA was synthesised. Like many other viruses, the hereditary material of the Semliki Forest virus – its genome – consists of RNA, not DNA, as in human cells. The genome of the virus is encapsulated in a protein shell, the nucleocapsid, which, in turn, is enveloped by a membrane.

Ossi Renkonen had also returned to the Sero-Bacteriology department, where he was my neighbour. Leevi and I tried to persuade him to join the team. He was a lipid expert, and we wanted to study membrane lipids in our projects. Ossi needed some time to consider our proposition, and I had to finish my earlier work. Our project would not quite get off the ground.

***

At this point, I had the opportunity to go to Cambridge in the UK to visit Staffan Magnusson and learn more protein methodology. Staffan was a visiting researcher in Brian Hartley's group at the Laboratory of Molecular Biology (LMB), a legendary stronghold of molecular biology. No fewer than twelve Nobel laureates made their breakthrough discoveries there. Francis Crick and James Watson found the double-helical structure of DNA; Fred Sanger sequenced proteins and developed a method to sequence DNA; John Kendrew and Max Perutz solved the first atomic structures of proteins, and Sydney Brenner contributed ground-breaking discoveries in several fields. These are just a few of the most famous scientists who worked at the LMB.

I was fascinated to see the daily routines there. Every morning at 10 am, there was coffee, then lunch at 1 pm, and afternoon tea at 4 pm. The dining room was furnished with long tables that filled up in the order that people arrived. This meant that your table companion might equally well be a technician or a Nobel laureate. It was incredibly democratic and interesting to meet so many different personalities. I might have an intense discussion about science on Monday and chat about the weather on Tuesday. The relaxed atmosphere favoured communication and surely inspired the researchers in their work.

Brian Hartley was friendly and made time to mentor me, even though I was there for only one month. His office was so tiny that only one other person at a time could fit in – similar conditions were shared by all the illustrious names at the LMB. Hartley came in to work late and often disappeared early. For an English top scientist, it was important to give the impression of not working. Supposedly, those who were intelligent and creative enough would be successful with little effort. Naturally, they had offices at home too. Today, we would say that they were 'working from home'. Now that this amazingly light touch has disappeared from England, I recall the attitude with nostalgia.

Brian Hartley and I took a train together to London to attend a lecture. During the ride, Brian wanted to hear about my research down to the tiniest detail. I started telling him about my serum proteins, but after a while, he looked at me – like my father – and asked "Kai, do you think that your research is interesting?" I was surprised by his frankness but admitted that, regrettably, it was indeed somewhat boring, and Hartley answered "You won't get very far as a researcher if you go on with those problems. Don't you have anything more interesting up your sleeve?"

Yes, indeed, I had. I described the Semliki Forest virus project to Brian and told him about the virus membrane, the world's simplest membrane model. I explained how the virus snatches a membrane envelope as it buds through the plasma membrane of the host cell. In the course of budding, the virus modifies the part of the cell membrane with which it is in contact, so the membrane proteins of the cell are pushed aside and replaced by those of the virus itself. The virus membrane thus becomes simplified during maturation to contain only one type of membrane protein. Hartley was fascinated by the process by which the virus wraps the cell membrane around its nucleocapsid and he encouraged me: "Kai, go for it!" He thought that the Semliki Forest virus was an attractive model for membrane research.

A weight was lifted off my chest. This was the jolt I needed to wake up – again. In fact, I had already understood in New York that this was the research I should engage in. Yes, my serum proteins lacked lustre. If an experienced scientist like Hartley thought that my research was more or less worthless, I was wasting my time. This had to change.

On the London train that day, I suddenly understood what Albert Szent-Györgyi had meant in a book I had read recently. Szent-Györgyi came from Hungary and belonged to the remarkable group of scientists who had received their education there before World War II. Several of the world's most brilliant physicists and mathematicians were Hungarians: Leo Szilard, Eugene Wigner, Edward Teller, John von Neumann, and George de Hevesy. Many of them were Jews who had to flee from the Germans. Albert Szent-Györgyi discovered vitamin C. He loved fishing and spent a lot of time at the Woods Hole marine biology laboratory on the Atlantic coast of the USA near Boston. When angling, he used really big hooks and had no luck. When asked why he stubbornly persisted with his gigantic hooks, he retorted that he was only interested in big fish.

In Cambridge. I stayed with a renowned physicist, Otto Frisch, and his wife. My father Lennart had worked with Frisch before the war at

Niels Bohr's institute in Copenhagen. Those were exciting times in nuclear physics. In Berlin, Otto Hahn and Fritz Strassmann had shown that when uranium nuclei are radiated with neutrons, a radioactive barium isotope is produced. Hahn could not understand how this was possible and corresponded with Lise Meitner in Stockholm. They had previously worked together in Berlin on these experiments, but since Meitner was Jewish, she had had to make a hasty escape to Stockholm in July 1938.

Over Christmas that year, Frisch visited Meitner, who was his aunt. During a walk in the woods with Frisch, Meitner devised an explanation for the puzzling result. She suggested that when uranium nuclei are pelted with neutrons, the uranium is split into barium and manganese. This liberates 200 million electron volts of energy per nucleus – according to Einstein's famous formula $E = mc^2$. What Hahn had observed was nuclear fission!

This was an enormous discovery, the basis of nuclear weapons as well as nuclear power plants. It is especially amazing how simple the technical equipment was that enabled Hahn, Meitner, and Strassmann to discover nuclear fission. It is on display in the Deutsches Museum in Munich, Germany. It fits onto a single lab bench and looks no more complicated than any bench in an ordinary electronics workshop. Today, nuclear physics is done on a completely different scale, using gigantic particle accelerators to reveal the secrets of elementary particles. The research projects of the European Organization for Nuclear Research, known as CERN, in Geneva, Switzerland, cost billions and are constantly increasing in complexity.

Upon his return to Copenhagen, Frisch performed a groundbreaking experiment based on the calculations Meitner and he had done in Stockholm. The experiment used a different method to demonstrate nuclear fission. My father was there and experienced the excitement that these results caused. It was stories like this that had inspired me to consider physics as a career. In fact, my father helped

Frisch to perform his experiment by building the measuring equipment that Frisch needed.

At the time of my visit to Cambridge, Otto Frisch was a very modest, elderly gentleman living in an ice-cold house without central heating. He confirmed the impression I got from Cambridge that top scientists often are unpretentious and kind individuals; however, when discussing mediocre research, they catch fire. It was drummed into me emphatically that mediocrity is the enemy of all research. One must focus on problems that are worth solving. The question is how to know which problems belong to this category. They also have to be soluble. The intellectual atmosphere in Cambridge was a revelation to me. I realised that I needed international contacts like these to stay well-informed. Above all, I understood that my expectations had been too low. If I wanted to succeed in science, I had to raise the bar for my scientific ambitions without getting cold feet. No risk, no fun.

***

So, I went to work. First, I persuaded Henrik Garoff to give up the Lp(a) protein – which he intended to be the subject of his dissertation – and, instead, to concentrate on the membrane of the Semliki Forest virus. This he did at first reluctantly, but then he remained true to this virus until his retirement. Henrik built his entire career on the Semliki Forest virus. Ari Helenius had previously used plasma lipoproteins to analyse how detergents – like those we use to wash greasy dishes – dissolve lipids from membrane proteins. Now, the virus membrane became Ari's greasy dish! Ari and I managed to understand how various types of detergents separate membrane proteins and lipids. The experiments provided a new perspective on the functions of detergents, which could also be applied to other cell membranes. Our articles, which became classics, still help researchers to deal with poorly soluble membrane proteins.

At the molecular level, biological mechanisms throughout the living world are almost identical. Life on Earth began with a primordial cell that formed between three and four billion years ago and has given rise to all organisms through the process of evolution. The most striking evidence for this is that we all 'speak' the same genetic language. This language is inscribed in every genome in every organism, including bacteria and even viruses, even though these are not, strictly speaking, organisms. The genetic language was deciphered by studies of the intestinal bacterium *Escherichia coli* and its viruses. The famous molecular biologist Jacques Monod formulated the maxim "What is true of *E. coli* is also true of elephants." All living beings on Earth are our relatives. This insight still has not really penetrated the public's consciousness.

This universality of life permeates all of biology, and it also applies to the molecular design of cells and tissues. If we can understand the mechanisms of molecular processes in an amoeba, a unicellular organism, we can be sure that our findings can be extrapolated to other cell types as well. This implies that a molecular biologist should search for the simplest possible experimental model in which to study the chosen process. As I was interested in the structure and assembly of biological membranes, which are crucially important for cell function, I had to find a suitable model membrane in which to do experiments. Cell membranes are immensely complex lipid structures with hundreds of proteins. They were too difficult to study with the methods available at that time, and they did not attract young researchers interested in the exciting research themes in molecular biology. They were felt to be too risky.

Fortunately, Brian Hartley was enthusiastic when I told him about the membrane of the Semliki Forest virus, which was so simple in its structure that it might reveal its secrets to us. A layman would certainly not have understood why we wanted to study the membrane of a virus rather than a proper cell membrane. But I was now convinced that, to paraphrase Monod, "What is true of the Semliki Forest virus membrane is also true of the cell membranes in elephants." I just followed the brilliant message of molecular biology!

My self-confidence had grown during my short stay at the LMB, and I felt like an explorer entering unknown territory, anticipating future discoveries. For me, it was important that if all cell membranes have the same general structure, then my findings with the virus membrane would be generally relevant. Membranes are extremely thin, essentially two-dimensional fluid sheets consisting of a double layer of lipids only five nanometres (five-thousandths of a millimetre) thick with proteins floating in it. Protein windows in the membrane open and close. They control which signals pass from the outside to regulate what happens in the cytoplasm and the organelles. These fascinating, thin lipid sheets are hives of activity. Thirty per cent of all proteins are membrane proteins, and a large proportion of all cell functions occur at or in the membranes themselves. They are wonders of biological design, truly worth exploring.

The Semliki Forest virus was an interesting research object in itself, of course. Many of the insights from our investigations turned out to be relevant to other viruses too. For example, the coronavirus SARS-CoV-2 acquires its membrane envelope in the same way as the Semliki Forest virus – by budding through a cell membrane.

***

Now the collaboration with Leevi's group really gathered momentum. Fortunately, we managed to convince Ossi Renkonen to join us. My dream from my student days had become a reality: I was now part of a troika with different skill sets. Together, we studied the secrets of the Semliki Forest virus. Leevi was a virologist; his group knew how to grow the virus in cells in culture and how to isolate it in milligram quantities. Ossi was a lipid expert; membranes contain both lipids and proteins. My own group concentrated on the structure of the membrane and the proteins it contained. Our troika studied membranes in the context of the host cell. Ours was the only lab at that time that could analyse both lipids and proteins in biological membranes.

My task became, by and large, to ensure that the collaboration worked smoothly, with the support of Leevi and Ossi, of course. It required continuous communication. My ill-tempered laboratory assistant Hilkka reproached me all the time "All you do is go around talking!" But collaborations are always running into obstacles. I wanted to follow what the scientists of our troika were doing and try to help them to realise their goals. I couldn't just sit in my room doing my own experiments and minding my own business. I was constantly on the move, visiting the laboratories of Leevi and Ossi and discussing the project with the researchers involved in our collaborative effort.

Our joint seminars every Monday morning became an important uniting element and, for us at least, legendary. All the members of our three groups, young and old, participated. The discussions were conducted in English, the lingua franca of science. We reviewed our work in detail, critiqued it honestly, and had animated discussions. We also analysed the literature to find out what was really known and to define the limits of our knowledge. We started to understand that our knowledge about membranes was very limited indeed. What was so fantastic about the Monday meetings was everyone's curiosity and the will to drill deeper and deeper. It may seem an exaggeration to say that the atmosphere was inspired, but during my long life in science, I have never again experienced such informative discussions. We took nothing for granted, and no one needed to feel ashamed of his or her ignorance. Christian Ehnholm had associated with my group to work on lipoproteins. He became an anchor of these meetings because he knew the art of asking simple but pointed questions that opened new perspectives. I realised later what an important role minds such as his play in stimulating creative thoughts.

One reason our seminars were so gratifying was that we felt like pioneers, like a collective of explorers. The fact that we attacked our research problem from three different angles – biosynthesis, the role of lipids, and the role of proteins – gave us more weight. And the fact that our viral membrane model was so simple prevented us from getting lost in the complex world of the real cell membrane. The simpler system was so much more accessible for experiments.

***

The Haartman Institute was a conglomerate of departments, each led by a tenured professor. Besides Sero-Bacteriology and Virology, it included three Pathology departments and Hygiene. Although all of them were in the same building, there were no interactions between the departments. To Leevi and me, this meant that there was no added value in being located under the same roof. Could that be changed? We wanted to initiate strategies that would promote contact between the departments and lead to collaborative projects.

First, we turned our attention to the institute's canteen, where the food was uninviting. In New York, we had enjoyed the hamburgers that were grilled in a bar in the basement of Rockefeller University. We persuaded the chef to let us grill hamburgers for lunch. We did that for a few days, hoping that the canteen would adopt our recipe. Unfortunately, this did not happen. The canteen staff followed our recipe for a week but then reverted to their old practice of picking up hamburgers from big tin cans. The headquarters probably thought our hamburgers were too expensive. We gave up.

We tried a different tack and went to talk to Professor Harald Teir. He was the leading figure who had secured political support and funding for the huge building project that became the Haartman Institute. Now, he was director of the entire institute, but his job mainly entailed handling communication with the university administration. I obtained my audience with Teir by telling Pappa

Renkonen that we would like to have more contacts across institutional borders to promote collaboration and communication between the researchers. Pappa Renkonen was positive, as usual, and conveyed my message to Teir. I proposed to Teir that he should appoint an unofficial board for the institute, with all department chairs as members. I volunteered to act as secretary of the board. Teir agreed and promised his support. Our first initiative was to organise a lecture series for the whole institute in our big lecture hall. The series became a success and contributed to making our Haartman Institute well-known in the biomedical research community and definitely improved communication within the institute.

Next, we proposed to set up a mechanical workshop to serve the researchers. This initiative was also approved and became an important support for our research. Many projects needed all kinds of small gadgets that the workshop could manufacture according to our designs. We recruited the trainer of the Finnish Archery Association, Veikko Virta, as the mechanic. He was good at making bows, so we thought he would have other mechanical skills suited to our needs. In fact he did not, but, fortunately, he was a natural talent and learned to make almost anything we needed. Before long, he was even selling our little gadgets to research institutions abroad. The workshop became a place where it was easy to meet and have informal discussions. Such small hubs are important in a research institution.

We also invested in social activities and founded a sports club that organised football training in the summer and ice hockey games in the winter. Although I was mediocre at both sports, I, too, took part and enjoyed every bit of it.

I learned much from Harald Teir. He was an experienced leader who knew how to handle difficult situations. The Haartman Institute had a budget for the maintenance of the building. Once, I was present when Teir negotiated with the finance director of the University about the following year's budget. The discussion rolled back and forth, but

Teir made no headway with his request for an increase in funding. Suddenly, Teir started shouting and repeated his request in an aggressive tone. Strangely enough, the finance director yielded. When we left, Teir told me "Kai, remember, this tactic can be used only once with the same person. Your counterpart gets so perplexed that he often agrees." I have used Teir's trick once or twice with success.

One anecdote from that time sticks with me and never fails to bring a smile to my face. A Soviet research delegation was shown around the Haartman Institute. Teir, who was also chairman of the Finnish society for cultivating the sauna culture, *Saunaseura*, acted as host. Naturally, there was a sauna in the department. As we passed it, he started describing his sauna research and opened the door. Inside, we surprised a janitor finishing a bottle of the Finnish vodka, Koskenkorva. Teir quickly shut the door, explaining that an important experiment was in progress that could not be disturbed. The Russians smiled understandingly.

***

The collaboration between the Semliki Forest virus troika was intensifying. We had secured a major research grant from the Association of Life Insurance Companies, which helped to consolidate our collective project.

As we were rather isolated in Helsinki, we realised that we had to do more to advertise our research on the virus. Without more effort, we would never get our articles published in good international journals. Many Finnish scientists published their findings in Finnish professional journals, which were read by practically no one. Our first Semliki Forest virus article was published in such a journal. We were never going to repeat that mistake; on the contrary, we advocated closing the domestic biomedical journals. They served no purpose anymore.

Ossi Renkonen managed to raise funding to organise an international conference on lipids and membranes. This gave us the opportunity to invite the leading researchers in the field to Helsinki. The conference was a great success and brought together a new mix of experts on membranes and viruses. Interdisciplinary meetings are generally most stimulating and generate new ideas, experiments, and, above all else, new contacts. It was equally important for us to present our own results and show what we had achieved. We took good care of our foreign guests, as was customary in Finland in those days. In our northern corner of the world, we did not have many visitors, but we wanted those who did visit to leave with positive impressions.

One of our foreign speakers was Vittorio Luzzati, a structural biologist who specialised in lipids. I had visited him once in Paris on a short study trip. He had shown that the lipids in cell membranes form not only two-dimensional layers but also highly plastic, three-dimensional structures. Luzzati's discoveries widened my perspective of what lipids can do and made me even more enthusiastic about membrane research. They implied that lipid membranes might play an important role in generating cell architecture.

Luzzati belonged to the elite of molecular biology, and he knew the structural biologists at the LMB. He was a friend of Rosalind Franklin, whose data were essential for Crick and Watson to develop their double-helix model of DNA. These two arrogant scientists published their revolutionary model without acknowledging the contribution of Franklin. During our conference in Helsinki, Luzzati stayed in our home. As an experienced traveller, he had brought with him a black eye mask to help him sleep through the light Finnish summer nights. Our children were awed by his eye mask.

After the conference, Ari Helenius was invited to Luzzati's lab. Luzzati wanted to add proteins to his research repertoire and needed Ari's help to learn to isolate poorly soluble membrane proteins using our detergent methods.

In my own lab, research progressed. We managed to figure out how the virus exits the host cell. Mark Bretscher from the LMB, whom I had met at a conference, had devised a new method for analysing membrane proteins and, in 1971, demonstrated that the major red blood cell membrane protein is a transmembrane protein. Amazingly, this was the first time a protein had been shown to span the lipid membrane; previously, scientists thought that membrane proteins were located on either side of the membrane. It shows just how little we knew about cell membrane structure at that time. My student Carl Gahmberg went to Cambridge and learned the new method of Bretscher's lab. Using this method, Henrik Garoff was able to demonstrate that the membrane protein of the Semliki Forest virus also traverses the lipid membrane and binds to the protein shell inside that surrounds the RNA genome – the nucleocapsid. Our data showed that each of the virus transmembrane proteins binds to one underlying protein in the nucleocapsid. It may sound trivial today, but thanks to this discovery, we could explain on a molecular level how the virus envelops itself in the plasma membrane of the cell.

I had read a new review of the research on this subject written by a scientist at Rockefeller University. He presented seven potential explanations of how membrane viruses acquire their envelopes and remodel patches of the cell membrane so they exclude host proteins. None of the explanations included our mechanism. Now we had data that would surprise the field of membrane biology. Our mechanism was the only one that explained how membrane budding could produce the virus envelope.

We submitted our article for publication in Nature, the most prestigious journal in our field. We waited and waited for a decision, only to be informed, at last, that the article had been rejected. We did not give up. We did not want to publish the mechanism in a specialised virology journal because then it would be read only by virologists. We thought it deserved a better fate. I wrote to Bretscher and asked him to contact James Watson to get the article published in the Proceedings of the National Academy of Sciences USA, which was next on our preferred journal list. Bretscher agreed, and Watson gave his OK. For a while, we felt like kings.

Now, I thought that I was qualified for a docentship (like a lectureship) in molecular biology at Helsinki University, but my application met with difficulties. Johan Järnefelt, the Chair of the Department of Medical Chemistry, did not regard molecular biology as an independent discipline. It turned into a typical academic dispute. Järnefelt classified molecular biology as biochemistry because that was his field. The problem lay deeper, however. The Medical Chemistry department was on the decline in Helsinki, whilst the Haartman Institute was becoming the leading centre for new talent. Besides the Semliki Forest virus troika, several other research groups at the institute – including those of Olli Mäkelä, Erkki Ruoslahti, and Antti Vaheri – were doing exciting research. Järnefelt had done groundbreaking research before he returned to Helsinki as a professor. He did not, unfortunately, manage to lead his institution to new heights after the retirement of his predecessor, Simola.

I remember one bizarre discussion I had with Järnefelt. He described himself as a scientist who took three months of holiday every summer on his island in the Gulf of Finland. There, he had time to think and bring depth to his research. I listened, astonished, but thought "Fine, but this is not reflected in what he had achieved in the past few years?" Järnefelt characterised me, somewhat condescendingly, as a scientific worker (*tieteentekijä* in Finnish, someone who 'makes science'), whereas he considered himself a scholar. Looking back at our clash now, I understand better what he meant; I did experiments with my hands, whereas he thought and philosophised. He saw me mainly as an angry young intruder into his domain. Despite his objections, I was awarded my docentship.

As a docent, I acted as deputy for the Professor of Medical Chemistry responsible for teaching in Swedish. It was the first time I worked as an academic teacher. When I gave my first lecture, loud laughter could be heard here and there in the audience. It was incredibly embarrassing; I did not understand what I was doing wrong. Later, I understood that searching for the right words made me stutter. I do not usually stutter, but it happens sometimes when I am poorly prepared to speak in public. Next time, I did my homework and happily, my lecture was much better. I have been spared similar debacles, but I have never quite lost the stutter.

The climate in the Finnish research community was becoming tougher. There simply was not enough government funding to support the university expansion that had occurred in Finland. Several new universities had been founded: Oulu in 1958, Kuopio in 1966, Joensuu in 1969, and Tampere University of Technology in 1965. Although it was established after World War II, the Academy of Finland assumed its present function only in 1970. Nowadays, the Academy is a central government agency that distributes research grants at several levels and across all disciplines. It comprises several Research Commissions, one for each major academic field, each with its own budget. In the early days, applications were peer reviewed mainly by Finnish experts, but now, since around 2000, applications in science and medicine are evaluated almost exclusively by foreign experts. The Medical Research Commission, which had enabled me to set up my own group at Sero-Bacteriology, was incorporated into the Academy of Finland. This reform of the funding system was a great improvement, but, unfortunately, the total budget was much too small – much smaller than in the leading industrial countries. The tight funding situation generated envy and unnecessary regional conflicts between researchers. I sensed that it would be hard to be competitive internationally unless our financial situation improved. Fortunately, several private foundations supported research. Without them, the situation would have been even worse.

***

Economic problems aside, the Semliki Forest virus troika was doing well. We were being recognised abroad and invited to international conferences. The greatest surprise for me was an invitation to a meeting in Heidelberg, Germany, focused on creating a new European Laboratory for Molecular Biology (EMBL) with Sir John Kendrew from the LMB in Cambridge as Director General. I went there with a feeling of awe and expectation. The European molecular biology elite was gathered to discuss the research themes of the new laboratory. Only four young investigators were present, including me. Many big names presented projects from their own fields. The plan was for EMBL to assemble and build the advanced instrumentation that they expected would catalyse progress in molecular biology. The model was the European Centre for Nuclear Physics (CERN) in Geneva, Switzerland.

I soon understood why I had been invited to Heidelberg: Vittorio Luzzati was there too. As a member of the council that advised Kendrew on planning EMBL, he asked me whether I might be interested in a group leader position at the new laboratory. Of course I was! He advised me to talk with Sir John. I had hardly dared to dream about an opportunity like this. Sir John informed me that a workshop on cell membranes was being organised and I should expect an invitation.

This was big news to take home, although it was uncertain what the outcome would be. Soon, I returned to Heidelberg to attend the membrane workshop. This was convened by a dynamic Italian, Ernesto Carafoli, and Sir John was also present. As I listened to some 25 senior and junior participants present results from their labs, my initial optimism faded. With so many talented researchers present, I would hardly stand a chance.

A week or two later, however, I received a third invitation to Heidelberg, this time to meet Sir John. I was offered a position as a group leader at EMBL. This felt like a miracle. Who would have believed this was possible? Finland was not even a member of the consortium of European countries that would fund EMBL. The member states were mostly the same ones that funded CERN, the Mecca of nuclear physicists.

Naturally, I did not want to move alone to Heidelberg, and I wondered whether I would be able to convince Carola. To my great relief, she immediately understood what this chance meant for me and needed no persuasion. How lucky I was to have lovely Carola as my life partner! I did not have to negotiate with our children, Katja and Mikael, as they were only four years old. And a third child was on its way.

Now came the worst part: breaking the news to Leevi and Ossi. Our project had been so successful, and our collaboration had worked so well. Yet, this was an offer I could not refuse. My companions were sad, but they understood that this was a once-in-a-lifetime opportunity.

I quickly realised that if I could convince Ari and Henrik to join me at EMBL, we would have a dream team. The salary was excellent, and the contract ran for three years. Happily, both realised what a unique opportunity it would be to participate in building a new European laboratory. Still, we all intended to return to Finland. I was glad that Hilkka also came on board when I asked her. She knew all our methods, so with her in the lab, things would work. The core team was ready. We were prepared to take on our greatest challenge so far.

Right in the middle of the preparations, I started to feel a strange pain radiating from my head. I had no idea what it could be. As a physician, I immediately feared a brain tumour – we always think of the worst possibility first. I still remember how I felt in Heidelberg, where I had gone to negotiate with the EMBL administration about the move and how our lab should be equipped. I lay crying in the bathtub in my hotel room. Life was smiling, everything was perfect. Except that I had a brain tumour and would die soon. Damn.

Back in Helsinki, I went to the university hospital. Jorma Palo, a neurologist whom I knew, examined me carefully. When he was finished, he asked "Kai, what are you doing these days?" I told him about my pending move to Heidelberg and what a stress it was to interrupt our activities in Helsinki and start from scratch in Heidelberg. Jorma looked at me calmly and responded "Kai, you are well. Your symptoms are typical stress symptoms." He didn't have to say more. I knew I was saved.

Chapter 4: Heidelberg

Our third child, Matias, was born, and we moved to Heidelberg quickly and seamlessly. The furniture was sent off, and I left with the twins. We had rented a house for two families, and Ari and Majlen had already moved in. As soon as I had installed everything in our half, Carola came with baby Matias and her mother. It was a warm and sunny day in May. We drove along the Neckar River into town, and the road was bordered by fruit trees in bloom. The old town and the famous castle were on the other side of the river. Heidelberg showed its best side. We thought that this boded well for our new adventure.

It was 1975 when we arrived; construction of the new EMBL building had not even started, so the researchers were accommodated in temporary premises. Our group – Ari, Henrik, Hilkka, and I would work in the newly built German Cancer Research Institute, the DKFZ, which also housed the EMBL administration led by Sir John Kendrew. EMBL had only thirty to forty employees. Thus, we were part of the venture from the very beginning.

EMBL had started its activities one year earlier. The political struggle about its location had been tough. The city of Nice, on the Cote d'Azur of France, had long been the favourite. Vittorio Luzzati recounted how the choice finally fell on Heidelberg. The renowned molecular biologist François Jacob had lunch with the French Foreign Minister, Couve de Murville. Many of the French elite knew one another because they were educated at the same schools. Couve de Murville confessed to accepting a decision that may not please Jacob. He had discussed CERN in Geneva and managed to make the German Foreign Minister concede an expansion of CERN onto French territory. The price was that Germany should have EMBL. Couve de Murville had asked how much EMBL would cost, and since it was much less expensive than the big CERN project, he agreed. The budget was more important than national pride. Naturally, Jacob was not pleased.

Why, then, Heidelberg and not Munich, as the Germans had initially intended? One reason was that the bus taking the EMBL delegation to the planned site in Martinsried had passed Dachau, the site of one of the most infamous Nazi concentration camps. The committee, including Luzzati, thought this association did not speak in favour of Munich. But another important factor was that Hermann Bujard, Peter von Sengbusch, and Ken Holmes had devised a compelling bid for Heidelberg presented by the City Mayor and had done a great job hosting the commission in Heidelberg. Kendrew and his advisors were taken to rather down-to-earth places, like the boisterous student pubs, in contrast to Munich, where they had been formally wined and dined. Kendrew liked the projected site in Heidelberg, near the centre but with a view over woods and fields from the top of a hill. I was also glad that Heidelberg had been selected. If EMBL had been located on the Mediterranean, many more accomplished candidates would have registered their interest, and I would not have stood a chance.

Our lab was on the seventh floor of the DKFZ. The rooms were empty, so Ari, who arrived first, had to procure all we needed to get started. For the position of administrative director, Sir John had recruited Jack Embling, who had been undersecretary to Margaret Thatcher when she was Minister of Education. Ari, Henrik, and I – our new troika – had composed a long wish-list of all the instruments and apparatus we needed. Ari handed the list to Embling and waited tensely for his reaction. Embling had a distinguished administrative career but had little notion of scientific research, especially its cost. Systematically, he questioned every item on the list. Every purchase had to be defended in detail in long discussions. Embling's philosophy seemed to be that if Ari was willing to take hours to explain what centrifuges and the other instruments on the list were, what one did with them, and why they could not possibly be substituted with something cheaper, then the equipment must be important and be purchased. Ari took Embling's questions seriously and passed the interrogations with flying colours. We got everything on our list except one machine. Here, Embling thought that his expertise was sufficient. The machine was a coffee maker that cost thirty Deutsche Marks. "If I accept this, all research groups will request a coffee machine!" he explained.

Some newly hired group leaders, especially the Germans, did not manage Embling quite so well but started quarrelling with him about what they thought was his irrelevant questioning. But Embling had no choice. He had no idea what all the instruments on the lists were needed for, and he used the occasion to expand his knowledge. At that time, EMBL still had a very light administration, and Embling had to make all decisions alone. Those who did not pass the tests found their lists slimmed down and had to detail their requests in writing to get them accepted.

An additional inconvenience for Ari was that EMBL still had no support staff. When the heavy instruments began to arrive, the goods reception at the DKFZ called Ari, who had to run down to help move the boxes up to the seventh floor. Procrastination was impossible. If he did not turn up within ten minutes, the reception threatened to send away the truck that had brought the equipment. The head of reception had been a chauffeur for the first director of the DKFZ and regarded his new position as a demotion. Ari and other foreigners were suitable victims to take it out on.

***

We moved into the two-family house in Ziegelhausen, a suburb of Heidelberg. Carola, Katja, Mikael, Matias, and I lived on the ground floor and Majlen, Ari, Jonne and Ira on the top floor. We had hired a

Finnish nanny, Arja, and carried on with the same system as in Helsinki. Majlen was a chemist and subsequently retrained as a librarian. She was employed by the DKFZ library. Half a year later, Carola started work as an associate with a dentist who had previously practiced in Sweden. German was a problem for her initially, but at least her boss knew some Swedish.

We introduced a system that made it easier for us to take care of the children. Each parent was allotted one day a week to come home early from work. On that day, you had to be at home at 4 pm and spend time with the children until 7 pm. It was important that all of us should have time to talk to Arja, whose workday ended at 5 pm. Soon Katja, Mikael and Jonne started to go to kindergarten, but they came home for lunch, which Arja prepared. When we asked the children how kindergarten was, all was fine. Were they picking up German? "No," they answered, "we don't need any German because the teachers understand Swedish." How unexpected! In reality, this was not true, of course, but our children formed such a close-knit group they were entirely self-contained. They did not need to communicate with the other children or the teachers.

On the first afternoon when I came home at 4 pm, I had planned to read a few research papers while the children were playing; but no, there were lively protests. "You shall have fun with us like Ari. *He* plays with us." After that, I had fun with them, too. It was quite a job to entertain five children. On the weekends, we all went out together. Beautiful woods, mountains and valleys surround Heidelberg. There was always some new place to discover. Sometimes there were hiccups with four parents and five children, but the issues that arose were easier to solve together than they would have been if we lived as two separate families. As we all lived under the same roof, we could have a nanny who was equally useful to both families.

Since we moved to Germany in the 1970s, we have seen a complete change in social attitudes. When we arrived, the position of women in Germany was quite different from that in Finland. As a rule, women in Germany, especially middle-class women, stopped working when they had children. They often remained stay-at-home mothers the whole time their children attended school. After such a long time at home, it was not easy to re-enter working life. The social pressures were heavy as well. A mother who worked outside the home by choice rather than economic necessity was called a *Rabenmutter* or 'raven mother'; in other words, a mother who does not care about her children. Kindergarten usually had a two-hour break in the middle of the day when the children went home for lunch. The system was simply not geared to the needs of working mothers. Mothers who had to work could, in principle, place their children in public day-care all day, but there was a screaming scarcity of places. Fathers were almost never seen dropping off or picking up their children at the kindergarten.

In Germany today, both parents work, and household chores are shared in many families. The German school system is mainly unchanged, however. Primary school from age six to ten is the same for all. At the tender age of ten, it is decided which of the secondary school types – *Hauptschule, Realschule* or *Gymnasium* – the child is suitable for. Parents with higher education usually ensure their offspring get into *Gymnasium* – the only secondary school leading to university. Only 27% of children from non-academic families go to university, compared with 79% of children from academic families. Although Germany is otherwise a very democratic society, the school system still resembles what Finland abolished with the basic education reform of the 1970s. In the 1980s, Germany also experimented with comprehensive schools, but the attempts were doomed to fail because of the enormous pressure at all levels of society. Not even the social democrats dared to fight in earnest for reform.

***

In our new lab at the DKFZ, we were ready to start doing experiments. I had persuaded Kendrew to fund a laboratory assistant for Ari and Henrik. Now we were really a hard-hitting team.

The summer of 1975 was unusually hot. Our rooms were right under the roof of the building, where the temperature rose above 30°C. There was no air conditioning, which made it impossible to work. Experiments that we had done at 20°C in Helsinki gave completely different results in Heidelberg. Nothing worked. Really, we should have taken a holiday for the whole summer and avoided all the stress.

We started to rethink our research strategy. So far, we had concentrated on the virus. Now, we shifted our perspective to include the host cell as well. We wanted to use the virus as a tool to investigate how cells are organised. Viruses infect a host cell and turn it into a virus factory producing thousands of new virus particles. The RNA genome of the Semliki Forest virus, and many other viruses, encodes the proteins that replicate its genome, the nucleocapsid protein, and the membrane protein of the virus. That is all. To get into and out of the cell, the virus piggybacks on the host cell's machinery. Ari focused on how the virus enters the cell, whilst Henrik and I studied how the membrane protein is synthesised and finds its way to the plasma membrane, where new virus particles are assembled. We wanted to understand what happened in the cell on a molecular level. Until then, virologists had tried to describe viral infections in cells and tissues essentially by means of images taken with an electron microscope. We wanted to take a different road, although naturally, we also used the electron microscope as a tool when it was required.

Our new approach was to use molecular cell biology. That was an entirely new discipline, whose pioneers were Günter Blobel and George Palade. We wanted to replace descriptive electron microscope-based cell biology with a mechanistic, molecular understanding of the events that underlie the functioning of cells and tissues. EMBL was the perfect place for such an ambitious project.

All multicellular organisms – including plants, insects, and mammals like us – have their DNA genomes packed into a nucleus in the centre of each of their cells. The genetic information stored in the DNA is 'transcribed' in the nucleus into molecules of messenger RNA, which are dispatched out of the nucleus to the cytoplasm, where they are 'translated' to build the machinery of the cell – the proteins. The cytoplasm, which surrounds the nucleus, contains an elaborate network of membranes that perform various functions. Membrane proteins are synthesised and inserted directly into a network of membranes called the endoplasmic reticulum. From here, they travel to the Golgi apparatus, a membrane complex that packages them for transport to the plasma membrane and other cell membranes. In addition to these membranes involved in synthesising new membrane proteins, membrane-bound endosomes, and lysosomes take in and process proteins and other materials from the outside. The whole is surrounded by the plasma membrane, which acts as a barrier against the outside but, at the same time, is equipped to direct the traffic of molecules into and out of the cell. All eukaryotic cells are built on this principle.

We used the Semliki Forest virus as a tracker to reveal these traffic routes in the cell. The virus enters the cell from the outside through endosomes and delivers its RNA genome to the cytoplasm. All the virus proteins are synthesised there, and the membrane protein is transported through the endoplasmic reticulum and Golgi complex to the plasma membrane, where new virus particles assemble. Our task was to clarify how this traffic system works and find out how the virus gets into and out of the cell.

***

With a new research project and a young family, I had more than enough to be getting on with. Nevertheless, I couldn't resist taking an interest in getting EMBL off the ground. We were soon going to move into the new building, and it was unclear how we would fill it. Where would the researchers come from? Well-known scientists came and went, never to be seen again. When we arrived, there were only seven group leaders in place. Strangely, most were Germans. We were so few that when a visitor gave a lecture at the laboratory and we went out to a restaurant afterwards, we could all sit around a single table. Kendrew himself often participated. I wondered whether I should dare offer my assistance in finding suitable candidates. Cell biology doubtlessly needed reinforcements. In the end, I decided to proceed on my own. Without telling Kendrew, I wrote to researchers I had met at conferences. I described the EMBL project and told them that we were looking for young group leaders. Could they recommend some brilliant postdoctoral researchers who might be eligible?

Bernhard Dobberstein was the first to take the bait. He was a postdoc with Günter Blobel at Rockefeller University and, together with his advisor, he had recently published a ground-breaking article leading the way for molecular approaches to cell biology. After him came Graham Warren from Cambridge, whom I had met at a conference. Warren was a membrane biochemist and made a strong impression on me with his brilliant analyses of our field. Next came Daniel Louvard, a French investigator, whom I had interviewed for a postdoctoral grant before he went to La Jolla in California to work with Jonathan Singer, a leading researcher on cell membranes.

The procedure was the same each time. I invited the candidate to Heidelberg to give a lecture. If our Semliki Forest virus troika was convinced he was suitable for us and EMBL, I asked Kendrew's secretary if the director might be in Heidelberg next Saturday. If that was the case, he was usually available in his office in a portacabin near the building site of the new EMBL. I knocked on his door and described our candidate. Kendrew interviewed me in detail and, finally, I could inform the candidate that he was in good standing but needed to apply formally for a group leader position. All three – Dobberstein, Warren and Louvard – were hired. Most surprising was that Kendrew did not even ask me what I was up to when I proposed more candidates but let me proceed without comment. Thus, we built an international team of young investigators who were all working with cell membranes. We became the core of a dynamic team of molecular cell biologists, a new research field in Europe.

EMBL received a great deal of criticism during the building phase. Many established molecular biologists thought it was a waste of resources to set up a European research laboratory and that they could have used the money better themselves in their own countries. I remember a conference on the Greek island of Spetses in 1975, which gathered a great part of the molecular biology elite. One afternoon, there was a discussion about EMBL and as the only representative of the laboratory present, I became the object of their accusations. It was not easy to counter their swingeing criticisms in a convincing manner. Suddenly, I realised that we who now worked at EMBL would be subject to critical scrutiny by the European molecular biology community. If our research went well, it would mean success and good prospects. If not, my career would be difficult.

Thus, it was important to deliver. The collaboration with Ari and Henrik worked well, and Hilkka and I ensured that the Semliki Forest virus was produced in sufficient amounts for all our projects, including a project with the Structural Biology division to determine the structure of the virus. Most importantly, our cell biology projects were advancing. I was confident that our team would deliver.

The working atmosphere at EMBL was very supportive. Everyone knew one another. We celebrated Christmas, Carnival and summer parties together, including our partners and children. The working language of the laboratory was English, but many of the support staff spoke only German. I suggested that we all use the informal '*du*' for 'you' when speaking German rather than the formal '*Sie*'. Selecting the correct form was just too difficult for us foreigners. This was not accepted very easily. Even some young Germans had been brought up so formally that they could not adapt to such a radical proposal. I also noticed that German janitors and technicians at EMBL thought that if we said *du* to them rather than the more authoritative *Sie*, they might, for instance, disappear early from work and take other liberties. We had to correct them and explain that we did it to strengthen the community spirit and make communication easier, but in no way did it mean that we should not perform our work as well as we could.

Sir John Kendrew attended the parties and talked to everyone until the new building was finished in 1978. After that, we did not see much of him. He stayed in his office when he was in Heidelberg and did not visit the labs. Strangely, I had lively and open conversations with him when we were one-to-one, but he became reserved when more than two people were present. Although he was very influential and knew half the world, he was shy. But EMBL was indeed his creation. Without him, the laboratory would never have gotten off the ground.

Occasionally, Sir John took me to exclusive meetings. One of these was at Brandeis University near Boston. We gave 'chalk talks' about our projects; we were not allowed to project slides but were restricted to using chalk and a blackboard. The method made the occasion more informal, and the purpose was to stimulate discussion. Sydney Brenner was a speaker. His talk was about the development of the experimental organism for which he would later be awarded the Nobel Prize, the nematode *Caenorhabditis elegans*. His command of English was exquisite, and he was sparkling with ideas. He was regarded as one of the most original researchers of his generation. Now, he was busy mapping the nematode's nervous system, which has only 302 nerve cells, a simple brain model. (Our brain contains more than eighty billion nerve cells.) Brenner wanted to decipher the wiring diagram of the nerve cells. He planned to make mutations that would disturb the functions of the nervous system. He hoped that he would then see how the connections between the nerve cells had been modified in the mutants and thus identify the code governing the system. It was an incredibly ambitious project carried out with simple methods.

After the lecture, I managed to ask a question. I asked him if he truly believed one could understand how the nervous system works by identifying the contacts between the nerve cells. Obviously, nerve cells do not only consist of contacts; their internal organisation must also play an important role. "Kai, you have a point," he answered. I almost burst with pride. But I did think that this approach was reductionism carried too far.

***

Kendrew never informed us about his own strategy for recruiting researchers to EMBL. Still, it was obvious that he was looking for young talent and avoided hiring established scientists. Occasionally, Sir John asked us for advice. I especially remember one case. He asked Ari and me to interview Karl Illmensee, an embryologist who studied how the mouse egg develops into an embryo after fertilisation. Illmensee was already a well-known name in his field. We were not so sure that he would strengthen EMBL. He replied so evasively to our questions and did not feel like a guy with whom you would enjoy having a discussion. When Sir John asked what we thought, we were hesitant. Illmensee didn't get an offer from Kendrew. That was lucky for EMBL.

Instead, Illmensee got a professorship in Geneva, where he assembled a major research group. In 1981, he published an article that made him world-famous. He demonstrated, for the first time, that it was possible to clone mammals by transferring the nucleus of a cell from a mouse embryo into a fertilised oocyte from which the nucleus had been removed. The study was a sensation. Soon there were rumours, however, that the results were flawed. Several postdocs who had come to his lab specifically to learn the technique were never allowed to see how the procedure worked. One of them told me that Illmensee did his experiments at night together with his wife. Something was wrong. The University of Geneva appointed a committee that investigated the case and concluded that there were problems. Yet he was not fired; he resigned only in 1988 when he obtained a new professorship in Salzburg. This case demonstrates how researchers sometimes use stubborn denial to navigate through criticism unscathed.

Young researchers who act as whistle-blowers often find it difficult to be heard. To me, this is unacceptable. Whistle-blowers must be respected. The problem of faking was often present in my mind as a research group leader. If someone in the lab were to fake data, would I find out? I was sensitised to the issue by two spectacular cases of fraud that were committed by William Summerlin and Mark Spector when they were postdocs. Summerlin worked in Robert Good's huge group of fifty researchers in the same institute where Leevi Kääriäinen had worked in New York. Together with his boss, Summerlin published an article demonstrating that it was possible to transplant skin between two genetically different mouse strains – from black mice to white mice – by cultivating the skin in a special medium. The article generated lively discussion at one of our Monday meetings in the Sero-Bacteriology department in Helsinki. If correct, it was an incredibly important discovery, but we were sceptical. Indeed, it turned out to be a case of fraud. Summerlin had simply dyed the skin black with his own hands. Good had accepted the fraudulent findings even though they were contrary to current knowledge. How could a person who was supposed to supervise fifty researchers have any idea what was really going on in his laboratory?

The Spector case was similar. He worked in the lab of the well-known scientist Ephraim Racker, and they published six articles together in top journals. At that time, I went to give a talk at the Massachusetts Institute of Technology, hosted by the Nobel laureate David Baltimore. David told me that Mark Spector had recently visited and done experiments in his lab. Spector had asked for a bed in the lab so that he could work at night. I found it strange that somebody needed to sleep in the lab, but David was impressed that Spector worked so hard. Later, it turned out that Spector also faked results. If his data had been correct, Racker would have profited tremendously, as they provided striking support for his important earlier work. These cases show how group leaders who are not critical enough can be fooled by faked results, especially if the data support their pet hypotheses. Wishful thinking is fatal for researchers; hubris can lead to their downfall.

David Baltimore himself ended up in trouble. He had collaborated with Thereza Imanishi-Kari, but the results they had published could not be reproduced by the graduate student who continued the work in Thereza's lab. I knew Thereza; she had worked in Sero-Bacteriology in Helsinki. She was a competent researcher and learned Finnish in no time. In reality, the affair was a trifle. Baltimore and Imanishi-Kari ought to have ensured that the results of the experiments in question were performed again with all the necessary controls. Baltimore refused. However, the ethical code of science requires that criticism is met with respect. If the results can be reproduced, all is OK. If not, the article must be retracted.

Now the USA's Federal Bureau of Investigation (FBI) entered the picture. A member of congress, John D. Dingell, had started a campaign to detect falsification in research. The FBI investigators studied the documentation of the experiments that had been done in Imanishi-Kari's lab and concluded that there was fraud. That was not true, however. If a similar team had studied our own experiments, they might have reached the same conclusion. Research data and notes are not annotated for detectives without experience in research. In industrial laboratories, all notes must be easy to interpret, and the protocols faultless. In academic research laboratories, where notes are taken only for the initiated, a certain degree of carelessness used to be acceptable. Today, this is changing. The standards have become stricter.

It took ten years for Baltimore and Imanishi-Kari to be acquitted. In the meantime, Baltimore became President of Rockefeller University but was forced to resign. Later, he became President of Caltech instead, another top address. If Imanishi-Kari and Baltimore had only repeated the crucial experiment, this whole commotion could have been avoided.

Our research culture has an inbuilt and unrelenting process of control against fraud, at least with respect to important data. If the results are untenable, they eventually disappear. This is also true of erroneous interpretations of good data; we must update our conclusions when new data change our earlier interpretations. Although research often takes two steps forward and one step back, it still proceeds slowly in the right direction. This is the process that drives the expansion of our knowledge base through research. Yet, the ethical foundation that should underlie the activities of every scientist requires continuous monitoring and needs to be stressed through teaching and training.

Those who run labs should not forget that fraud can occur anywhere. Once I was suspicious of the results of a postdoc who had a streak of experiments that all succeeded. This is very unusual, so, to me, it was a warning sign. Together with another group leader, I constructed an experiment to check the results of the experiments by electron microscopy. These were experiments where the ingredients were mixed in a test tube, they could be seen in the microscope and the outcome could be determined as well. Luckily, everything was fine.

***

I decided not to waste time learning German. After all, we had contracts for only three years. I tried to communicate in German using the words I remembered from school without being bothered about the grammar. Still, we subscribed to a German daily newspaper. First, I read only the headlines and translated them into Swedish. Then, I began to read the articles with a dictionary in my hand. I learned to read quickly, but speaking correctly was not so easily mastered. Ari decided to skip the problem with the articles *der*, *die,* and *das*. He claimed that *die* is the most common article, so he always used *die*. Over the years, my German naturally improved. The only way to learn a language is to speak it.

As a dentist, however, Carola had to speak perfect German. She took language courses and learned quite quickly to speak fluently. Her teachers also became friends with the family. One of them, Thomas Vogel, became one of our best friends. He sang songs, wrote his texts himself, and defended a doctoral thesis about French *chansons*. Thomas even made recordings and gave concerts, either accompanied by a band or alone with his guitar.

Carola and I occasionally went along to listen to Thomas' concerts. Sometimes he was fantastic; at other times, it was only so-so, and Thomas was unable to engage the audience. I have had the same experience as a lecturer. Even in science, you need a certain showmanship to capture the listeners' attention. You must modulate your voice, talk intelligibly, and show slides that are interesting and easy to understand. The same lecture may go well three times, and then, the fourth time, it is miserable. Then, it is best to wait for new results before trying again. It is simply a must for a scientist to be able to perform. Practice helps. With time, I almost got rid of my stutter and learned to articulate so the audience understood me.

We also went with Thomas to listen to concerts by other singers. He introduced us to Leonard Cohen in person when he was in Heidelberg. Once, Carola was present when Thomas invited home the famous East German singer–songwriter Wolf Biermann who was obliged to stay in West Germany after he gave a concert before the trade union *IG Metall* in Cologne in 1976. The leadership of the German Democratic Republic (GDR) would not allow him to return home. Thomas asked Biermann's opinion of his chances as a singer-songwriter or *Liedermacher*. Thomas sang and accompanied himself on the guitar while Biermann listened. When Thomas stopped, there was an embarrassing silence, after which Biermann shook his head and gave his honest answer. It was a tough disappointment for Thomas. Hard criticism is never pleasant. I was reminded of the time I told Brian Hartley about my mediocre research in Cambridge. Deep down, Thomas knew that he was not going to be a star. Instead, he became culture editor and later Head of Culture at the radio station *Südwestfunk*. He was a success in his new job, and maybe later he was pleased that he had asked Biermann for advice, as I was after talking to Hartley.

When Wolf Biermann was exiled, some of the most well-known Finnish musicians and singers sent a letter to the political leader of the GDR, Erich Honecker, congratulating him for getting rid of an enemy of the state. I was shaken when I saw the news in German papers, thinking they were shamefully naïve. The old-school communist movement in Finland, known as Taistoism, culminated at the end of the 1970s. This Stalinist minority fraction of the Communist Party attracted intellectuals of all ages in Finland. Strangely enough, it became very influential in Finnish cultural life, which shows how isolated Finland was. Had the Taistoists never read Arthur Koestler's

'Darkness at Noon' or Manès Sperber's 'Like a Tear in the Ocean'? How could an influential movement in a country on the border between east and west so strongly side with the Soviet system? Western capitalism was by no means perfect either, but who would not rather live in the West than in a communist dictatorship? When the Soviet leader Nikita Khrushchev visited Finland in the late 1950s, the Finnish President, Urho Kekkonen, stated in a lunch speech that even if all other countries in Europe turned communist, Finland would remain capitalist. Finland was the only country bordering the Soviet Union that remained independent after World War II. It had to navigate carefully to stay on friendly terms with its large neighbour. Sometimes the Kremlin leadership had to be reminded of our status, though.

***

Sir John Kendrew supported our Semliki Forest virus project. I kept him posted about our progress but was never quite sure whether he was really listening. Still, I gradually understood that Kendrew had passed on what he had heard from me. One Christmas, EMBL even sent a Christmas card with a picture of Semliki Forest viruses descending like snowflakes over the Institute. The snowflakes were beautiful; their design was based on our electron microscopy pictures of the virus. Still, I was afraid that the card might be construed as an indication that we were handling our virus carelessly. At this time, there was a heated debate worldwide about the dangers of gene manipulation. The DNA revolution was underway, and the Heidelberg press speculated about monsters in the forest around EMBL, where we had started using the new technologies and were developing methods for DNA research.

As early as 1975, there was a conference at Asilomar in California, where the cream of molecular biology practitioners discussed how the community should handle the new methods that made it possible to take DNA from the genome of one organism and transfer it into another. Might these modified DNA molecules spread through the natural environment and cause problems? The molecular biologists felt that they could not estimate the risk, so they decided to introduce a moratorium on such experiments during which any possible dangers could be investigated.

Many critics completely misinterpreted this cautionary measure. Opponents of molecular biology could not grasp that the scientists working in this field did not know whether the research was risky. They thought that when scientists warned about potential dangers, it could only mean the use was risky. The critics had convinced themselves, without evidence, that DNA manipulation was dangerous and must be prohibited. The Asilomar meeting was organised to show the way to the future and demonstrate the research community's sense of responsibility for society. The molecular biologists wanted to avoid the mistakes of the nuclear physicists by being transparent about potential security problems associated with the new technology before any such problems might arise.

Several of the leading molecular biologists who had originally supported the moratorium regretted this measure later. They sacrificed years defending DNA technology even when it became clear that the risks were minimal and easy to control. The public did not trust the research community then, and the situation has only worsened today.

Why should scientists not inform the public of what they are doing? We must do much more to win the confidence of society at large by continuously informing politicians and the public not only about advances but also about possible consequences. We experienced this problem almost every day during the COVID-19 pandemic. Science delivered new vaccines and treatments, but this was not enough; the lines of communication between biomedical scientists, social scientists, and psychologists were poor, impeding the delivery of a consistent message to the public.

It would certainly be wrong to abandon a strategy of transparency based on Asilomar's negative experiences. Ultimately, openness is the only way to convince society about the necessity for research and innovation. Almost every ground-breaking discovery or technological innovation will also have some negative effects that are hard to predict. When they do appear, society must act. It is the duty of scientists to monitor developments and inform the public about what is happening.

We need new protective mechanisms that do not follow national borders. In Europe, the European Union (EU) is best placed to establish procedures regulating the innovation process so that industrial lobbyists and other interest groups cannot take control of public opinion. The industry knows that the best way to counteract scientific criticism of their products is by funding research showing the opposite. Brutally, and without concern for ethics, for-profit companies have funded fake science showing their products are harmless. This happened with tobacco, fossil fuels, fast foods, lemonades, and sugar. The tactic usually works.

The car manufacturer Volkswagen manipulated its own research results in order to demonstrate that the emission controls of its vehicles met USA standards. With no apparent pangs of conscience, the company management advertised their diesel cars as environmentally safe. How could Volkswagen commit such an outright fraud?

Protection of integrity in both research and technological innovation is a must in our ever-more technical world. For researchers and teachers, it is one of our most important social obligations that should be emphasised in training both researchers and engineers. We must do all we can to ban industry incentives that effectively bribe researchers to produce and publish faked or biased data. These practices have gotten out of hand.

***

In Heidelberg, so far, we had been doing basic research with no commercial implications; but I was thinking that our virus research might also be turned into something useful. I was expecting to return to Helsinki soon because I was not sure whether my three-year contract with EMBL would be prolonged. I knew I could not do basic research in Finland at the same level as at EMBL, so I initiated a vaccine development project exploiting our insights into how to use detergents to produce subviral particles of the Semliki Forest virus for vaccines. I hoped this would be my main research focus when I returned to Finland. I wanted to find the most effective way of presenting the virus membrane protein – the spike protein – the main antigen of the virus, to the immune system. We prepared three different forms: a spike protein monomer in detergent, micelles containing eight spike proteins in a water-soluble, detergent-free form, and what we called virosomes – spike proteins reconstituted in lipid bilayers. We vaccinated mice with the different spike protein preparations. The results were most encouraging. When immunised with low doses of the protein micelles or the virosomes, the mice were totally protected against the lethal encephalitis caused by the virus. The monomeric form of the protein was practically inactive. We patented the discovery and published the findings in Nature in 1978. The vaccine research field was not interested in our results, however, and the patent interested no one in the drug industry. Generally, pharmaceutical companies stayed away from vaccines because potential profits were too small. A few shots of a vaccine are enough to prevent disease, whereas drugs often must be taken for life. The industry's neglect of vaccine development was apparent every time we were threatened by a new virus such as the human immunodeficiency virus (HIV) or Ebola virus.

SARS-CoV-2, the coronavirus that causes COVID-19, was the exception to this sad rule. The companies that developed the new generation of successful RNA vaccines have earned a fortune. Also,

Novavax developed a protein subunit vaccine that was successful against SARS-CoV-2. This vaccine was constructed according to the principles we elaborated on over forty years ago. Indeed, the article that describes the vaccine cites our work. This warms my heart. Our findings were relevant, after all. We were just ahead of our time. The protein micelle vaccine can be stored in the refrigerator rather than in expensive freezers making it an ideal vaccine against COVID-19 in low-income countries. This experience exemplifies how long it can take before basic research leads to practical results. This message needs to be repeated over and over again. Basic research is a key driver of innovation, but it takes time.

***

I chose vaccines as the subject of my future research because Finland has a strong track record in this type of research. Kari Penttinen developed a vaccine against mumps, used to vaccinate conscripts in the 1960s, before a combination vaccine against measles, mumps, and rubella was introduced in 1982. Pirjo Mäkelä established vaccines that were effective against meningitis in children. In 1967, Harri Nevanlinna produced a Rhesus-globulin preparation that prevented jaundice when a mother forms antibodies against her baby. The Finnish Red Cross distributed the preparation over the whole country, and Rhesus incompatibility was eradicated as a health problem in Finland, the first country in the world with such a program.

I applied for a vacant chair in biochemistry at the University of Helsinki and got it. I could hardly hope for a better position in Finland. But, at the same time, things were going so well in Heidelberg that I was reluctant to leave my basic research at EMBL. Kendrew made me a counteroffer of a permanent position, even though all positions were supposed to be limited to nine years. The idea was always that EMBL would train us and that we should return to new tasks in our home countries when the contracts ended.

Nevertheless, I decided to reconnoitre the terrain in Helsinki to find out whether working at the Department of Biochemistry would be possible. If there were sufficient resources to continue the vaccine project, I would be willing to leave EMBL. But I found the situation in Finland in 1979 was hopeless. The Department had no modern equipment, and the university couldn't promise substantial new investments. I decided to decline the professorship. However, I did not want to retreat with my tail tucked between my legs. Instead, I wanted to broadcast how bad things were in Finland. I called a press conference that was reported in all the media. The response was so great that when the head of staff at EMBL picked up two Finnish hitchhikers in Belgium on his way to Heidelberg and mentioned that he knew a Finn, Kai Simons, both exclaimed that Simons was getting lots of attention in the Finnish media. That was my 'Andy Warhol moment' – this leader of the pop art movement claimed that in the future everyone would have fifteen minutes of fame.

My intention was not to denigrate Finland but to call for help. The situation of researchers was catastrophic. I worked as an expert reviewer evaluating the Department of Molecular Genetics at the University of Cologne in Germany. Its fifteen research groups had more money at their disposal than the budget of the entire Medical Commission of the Academy of Finland. The Academy was the main source of funding for basic biomedical research in Finland. Medical research was funded by the Medical Commission, to which several new areas, such as occupational safety and social medicine, had recently been attached. Obviously, these areas needed support, but when the budget did not grow, individual grants became smaller and smaller. From an international perspective, these grants were a drop in the ocean.

It was depressing. Finnish politicians thought the country could manage without research. They fantasised that Finland could buy the know-how it needed and did not understand why this could not possibly work. Pappa Renkonen gave me an example: the Finnish government founded a state-owned factory for producing antibiotics and bought both the patents and know-how from abroad. But who had the competence to make the right decisions? That competence did not exist in Finland. It was a complete failure, and the factory went bankrupt.

Finland needed its research base to create new products. My cry for help may have contributed to the change in the 1980s and 1990s, when Finland began investing heavily in research. In any case, my actions were strongly approved by the research community, suffering from the lack of resources. Normally, someone who declines a position becomes *persona non grata*, but, in this case, I was praised.

***

The Semliki Forest virus troika remained in Heidelberg, and we continued to work on our virus. Now, we wanted to join the DNA revolution. Henrik and I decided to sequence the genome of the virus to determine its genetic code. The genome is made of RNA rather than DNA, but RNA can be transcribed into DNA in the test tube, and that DNA copy can be sequenced to reveal the genetic code.

EMBL was busy developing methods for faster, more efficient sequencing. We heard that several other groups in the world intended to sequence the Semliki Forest virus or one of its cousins, so we had to act fast. I decided to serve as Henrik's technical assistant and sequence all the genome fragments he cloned. I also handled the new electrophoresis system for DNA sequencing that Wilhelm Ansorge at EMBL had developed, and I ran it without interruption day in, day out. Electrophoresis has been my specialty ever since my time at Rockefeller. Our next task was to assemble all the small pieces of sequences into a whole RNA molecule. Here too, we received help from another group led by Hans Lehrach, whose new method facilitated the task. We completed the sequence in record time, and it was one of the first RNA virus sequences to be reported. Our competitors were left far behind.

When we published the study, my friend Thomas Vogel wanted to see the article. He was disappointed to see my name in the middle of the list of authors. The work was a team effort, and it was only important for me to get the job done. Henrik was the first author, and for him, the article was a bright feather in his cap. I decided not to become an expert in recombinant DNA myself; the methodology was spreading rapidly, and it would be easy to engage researchers who were technically skilled in that field for my research.

We continued our studies to clarify how the virus enters the cell and how the newly produced viral particles exit the cell. Ari had shown that the virus binds to receptor proteins on the cell surface. Then the membrane invaginates to produce a vesicle that transports the virus to the endosome, a membrane-bound organelle in the cytoplasm. Ari also discovered how the virus gets out of the endosome. His work was spectacular and secured his future career. The membrane of the virus fuses with that of the endosome and – swish! – the nucleocapsid passes into the cytoplasm, where the viral RNA then instructs the host cell to produce many new RNA molecules and virus proteins.

The route used by the virus to enter the cell through endosomes normally ends in the lysosomes, which are intracellular vesicles filled with degrading enzymes – the garbage processing centres in our cells. Semliki Forest virus avoids destruction by opening a door in the endosome's membrane and escaping into the cytoplasm before it reaches the destructive lysosome. The virus membrane works like a stealth bomber, delivering the virus nucleocapsid into the cell without being detected. The cell succumbs when the RNA of the virus takes control but stays alive long enough for us to do our experiments.

Henrik demonstrated that the membrane protein of the virus is synthesised by ribosomes – the cell's protein factories – stuck to the endoplasmic reticulum. The newly synthesised virus membrane proteins exit the reticulum and are shuttled to the Golgi complex, where they are sorted into membrane packets, transported along protein cables to the plasma membrane, and assembled into new virus particles.

Our account of the life cycle of a membrane-enveloped virus was the first of its kind, and it was published in Scientific American under the title 'How an animal virus gets into and out of its host cell'. At home, we had always subscribed to this popular science journal, and I read it avidly. Now, our own article was published there; one of my dreams had come true! Readers of Scientific American worldwide would be able to learn about our research. We had succeeded in using viruses as tools to discover how RNA and proteins are transported into and out of the cell. Thereby, we had fulfilled the hopes invested in us. Our virus had proven a reliable reporter for understanding membrane trafficking in the cell. Our studies helped lay the foundation for a new chapter in molecular cell biology.

Ari was now ready to leave EMBL and was applying for independent positions. There was nothing worth having in Finland. At first, it looked like he might move to Sweden, where the conditions were significantly better, but George Palade from Yale intervened. He persuaded Ari to move across the Atlantic to New Haven. Ari had a phenomenal career. He later became chair of Palade's renowned Department of Cell Biology at Yale. Surprisingly, he was rejected by David Baltimore when he was looking for jobs in the USA. David scrutinised his *curriculum vitae* and remarked that he had stayed too long in my group. He did not understand that we worked as a team and that this had been the basis of our success.

David's reaction reveals one of the paradoxes of the research community. Cooperation is not rewarded. It is customary to identify a single person behind each study. There are almost always several authors of scientific articles in our research area, but the group leader usually gets the credit. This is how it was then, and the situation has not improved. Nevertheless, Ari got many offers, as many people knew his worth. After the second lecture on his US tour, he lost his slides and had to use chalk and blackboard at his future seminars. But Ari was such a skilled draughtsman that the talks were almost better with the live illustrator. He impressed the audience. I would have failed miserably in that situation.

***

In 1982, Sir John Kendrew had to resign. The European molecular biology establishment had never quite supported EMBL and was disappointed with the institution's achievements during Kendrew's reign. Not everyone understood his strategy of not recruiting established researchers but preferring a more open organisation with young group leaders. Kendrew knew that senior scientists would undoubtedly build their territories in the institute in traditional European style. Instead, he invested in young talent, who were given the opportunity to realise their own ideas without interference. No hierarchical barriers should hamper communication and collaboration. Thus, the development at EMBL resembled our modest attempts to stimulate research at the Haartman Institute in Helsinki.

In fact, Kendrew's track record from his reign at EMBL was more innovative than anyone could imagine in 1982. In retrospect, it became obvious that what EMBL had achieved under Kendrew exceeded all expectations. At the EMBL branch in Hamburg, Ken Holmes built a station that used the particle accelerator of DESY (the *Deutsche Elektronen-Synchrotron*) to analyse protein structures at the atomic level. It attracted biologists from Europe and the USA who wanted to determine protein structures that could not be solved by conventional means. The initiative was copied all over the world.

While working at EMBL, biophysicist Jacques Dubochet developed the vitrification method, essential for cryo-electron microscopy, earning him a Nobel Prize in 2017. Cryoelectron microscopy technique, which is used to determine the structures of proteins and other biomolecules at high resolution and in solution after quick and careful freezing, is revolutionising structural biology today.

Developmental biologists Christiane Nüsslein-Volhard and Erich Wieschaus also won a Nobel Prize in 1995 for their work on the developmental biology of the fruit fly, which was partly undertaken while they were investigators at EMBL. They discovered genes that control the formation of segments in early fly embryos and were pioneers of molecular developmental biology.

EMBL was a forerunner in molecular cell biology and raised the first generation of European researchers in this discipline, which became a mainstay of biomedical research worldwide. Through its investment in young investigators and promotion of collaboration under generous leadership, research at EMBL flourished. Compared with the static and hierarchical environments of most European research institutes at that time, EMBL was unique.

Before Sir John Kendrew left EMBL, the city of Heidelberg wanted to organise a farewell party in the castle. Kendrew was not at all interested. He did not like big celebrations and hated shaking hands. Diplomatic etiquette, unfortunately, required this tribute, and he was compelled to submit. EMBL had to cultivate its relations with its host city. When the great day arrived, Sir John greeted the guests at the entrance. But he shook no hands. His right arm was unavailable because he had wrapped it in a sling.

Manfred Eigen gave an interesting talk at the celebration. Eigen belonged to the senior German molecular biology elite, and he had a megalomanic vision for EMBL, which he thought should resemble CERN with gigantic instruments constructed by huge teams of technicians, engineers, and researchers. He imagined that all data from these instruments would be transferred to other European laboratories for analysis and interpretation. After that, super scientists such as himself would decipher the fundamental codes that regulate how life functions at the molecular level and make big discoveries.

In reality, EMBL turned out to be diametrically opposite to Eigen's vision. The researchers formed small groups that collaborated and complemented each other like the cells in an organism. These groups' combined activity resulted from a concerted effort and brought enormous added value to EMBL. Big science was replaced by synergistic small science. This collaboration was the driving force for research in cell biology at EMBL, but the freedom that was given to EMBL's young research group leaders was just as important.

Kendrew's successor as General Director of EMBL, Lennart Philipson, came from Sweden. He pioneered Swedish molecular biology at a time when biochemistry in that country, as in many other parts of Europe, dominated the biological research landscape.

Philipson was quite a different type than Kendrew. He circulated around the laboratory, exchanging a few words with everyone. He made no distinction between gardeners and group leaders. You would suddenly feel a hand on your shoulder, and there was Philipson. He was a big man with a bass voice and a beard. We called him the Viking. Also, he spoke quite differently from Kendrew, who was very careful in his choice of words. Kendrew pondered his decisions thoroughly, but once he had made a decision, he was almost unshakable. Philipson was more casual in his parlance. His first speech to the EMBL researchers terrified the British. They assumed that Philipson's plans were fixed, just as he presented them. In fact, they were not. If you had a different opinion, you could argue your point. If your arguments were convincing, Philipson would change his mind.

I knew Philipson from my time in Helsinki. He was a molecular virologist, like us, and was interested in the molecular machinery that synthesises viral genomes. Philipson wanted to reorganise EMBL. This was really needed because Kendrew had placed the research groups helter-skelter in the laboratory with no attention to the topics each group studied. Kendrew's idea had been to let random choice stimulate fruitful contacts across disciplinary boundaries. As EMBL grew, however, it became necessary to place those groups that used the same instruments close to one another. So, it also became easier to maintain contact between groups that were engaged with similar problems.

Philipson wanted to introduce a new structure. He created several thematic programmes, including Cell Biology, Instrumentation, and Structural Biology, and added a Cell Differentiation and Gene Expression division. I was appointed Director of the Cell Biology programme. To downplay the hierarchy, I suggested we should be called coordinators rather than directors. The core of our programme was already in place and consisted of the group leaders I had assisted in recruiting. We had to move from our locations dispersed throughout the building to one area of the laboratory where we were all together and could work even more efficiently as a collective.

***

I felt ready for a change of research theme. Now that we had clarified how the Semliki Forest virus enters and exits the host cell, it was time to find a new problem. I had a permanent position and wanted to raise the bar even higher by embarking on a voyage of discovery with new goals. Ari and Henrik could continue with the Semliki Forest virus for as long as they stayed at EMBL and take the projects with them when they left.

When Ari decided to move to Yale, our group was split in three. Henrik became a group leader in the Cell Biology programme. For myself, I kept only my assistant Hilkka and Karl Matlin, a new postdoc from Rockefeller. I wanted to go on using viruses as reporters in a cell type that would allow me to study cell polarity. The question of how some cells adopted a polarised structure was important for biology and had not yet been explained in molecular terms. After a period of fruitless searching, I finally found my dream cell type. Daniel Louvard had joined EMBL from the USA and brought with him a polar epithelial cell line, MDCK, derived from a dog kidney. We could culture these cells and use them as a model system for studying cell polarity.

Epithelial cells cover all the surfaces of the organs in our body. They are packed together tightly, sometimes in a single cell layer, sometimes in several layers, which function as a protective barrier to the outside world. The cells in each layer are polarised, meaning they have an apical pole facing outward into a cavity, such as the gut, and a basal pole facing inward towards the bloodstream. The poles have different functions because the cell's plasma membrane at each pole contains a distinct set of proteins. In the kidney, this enables selective transport of ions and molecules through the cell from one side to the other. In the intestine, the apical poles of the epithelial cells face the cavity with all its digestive juices, and the basal poles face the blood vessels, allowing the transport of nutrients from the food into circulation.

I chose to study MDCK epithelial cells after I read an article by cell biologists Enrique Rodriguez-Boulan and David Sabatini, who had shown that two membrane viruses (influenza virus and vesicular stomatitis virus, or VSV) are shed from these cells in a polarised manner. New influenza viruses are liberated from the apical pole, whereas VSV is released from the basal pole. We could study how virus membrane proteins traffic to the apical and basal poles in MDCK cells, an excellent model system that allowed us to continue using our virus reporters. Unfortunately, the Semliki Forest virus did not grow in MDCK cells, so we would have to learn to work with these two new viruses. Also, we would have to develop novel methods to discover how the cells generate their asymmetric architecture.

The Cell Biology programme took on the task of improving our methodological arsenal. My group was already introducing DNA methods into cell biology. Daniel Louvard was a master at producing antibodies, which would be needed for identifying and characterising proteins in cells and tissues. He initiated a project for producing monoclonal antibodies based on a new method from the LMB in Cambridge. And Gareth Griffiths was developing a cryoelectron microscopy method using antibodies to show the locations of proteins in cells and tissues.

Great breakthroughs in biology are almost always the result of new methodologies. We needed to widen our methodological repertoire to solve the problems we faced. We also wanted to disseminate the new methods we had learnt. To do so, we organised biennial methods courses at EMBL. These became so popular that, for a while, almost all successful cell biologists in Europe had attended at least one of them. The courses also attracted young talent from whom we recruited postdocs and group leaders who would go on to build the new molecular cell biology of Europe.

Gerrit van Meer from the Lipid Research Centre in Utrecht participated in our first cell biology course and became an important addition to my lab. He had not imagined himself becoming a career researcher. He thought we worked too hard; perhaps he envisaged a more pleasant life. I managed to persuade him, though, and he has probably never regretted his decision. We needed a lipid biochemist to extend our study of membrane protein trafficking in MDCK cells to include membrane lipids as well. No other group tried to find out how lipids and proteins come together to form the membrane that envelops the viral nucleocapsid.

The EMBL methods courses began with two days of lectures given by invited researchers. Systematically, we invited all the most important cell biologists in the world to our course. In this way, we got to know them, and they got to know us. The programme included a visit to a local new wine festival, which meant that the courses had to be held in September when the wine was fermented in the Heidelberg region. Gareth Griffiths was responsible for the organisation and entertainment. Spirits were high, and the wine was flowing. The next day was always somewhat foggy, but many participants still cherished their memories of our courses and parties.

Being a researcher is frustrating. A project does not always advance. Experiments fail. It is easy to get stuck. Then you must step back to make room for new thoughts. We all have our tricks, but I have noticed that a certain playfulness stimulates the imagination. Why should everything be so serious? Gareth was a good example. He was not only a skilful electron microscopist but a talented clown as well. We arranged parties where we put on amusing shows. We made fun of each other and had a good time. We wanted to remind ourselves that, although life is serious, a playful mood can work wonders. Pub crawls were also popular at EMBL. I participated once, but I did not do it again. It resulted in a terrible hangover.

***

My style as a leader was like that of Philipson. I walked around and discussed science with everyone. On the one hand, it was important for me that the group leaders in the Cell Biology programme found their own lines of research. On the other hand, to ensure that the doctoral students, postdocs, and group leaders could freely exchange ideas and get feedback from each other, those lines should not diverge too far from what the whole programme was doing. Occasionally, I sketched new projects and tried to entice group leaders to adopt them. I seldom succeeded. Sometimes the effect was quite the opposite, but that did not matter. My attempts stimulated counter-thoughts. I wanted to promote a discussion about synergistic strategies, especially for new group leaders, and generate creativity and collaboration. Graham Warren was a great asset. He helped to keep the critical discourse going. He dug deep into issues from all sides in that incisive way he was taught in Cambridge.

The Cell Biology group leaders held meetings where we discussed and argued without restraint, whether about recruitment, research projects, budgets, courses or equipment. If we did not reach a consensus, I decided. Remarkably, the others accepted this. By then, perhaps, they felt they had all had sufficient occasion to air their views and let off steam. The sessions put a heavy load on me as the moderator and could result in a blistering headache, but by the next day everything was fine again.

One of the most demanding tasks of a researcher is to keep up with new developments that affect your own field. You can learn a lot by attending lectures at your own institution, by going to conferences, or through the grapevine. Yet the most important source of information is original research articles, which you must read yourself. You should not rely on second-hand knowledge. Rumours are often distorted. Occasionally, you might hear something interesting presented in a lecture, but when you read the original article it turns out that the experimental results, in fact, lead to the opposite conclusion.

I have always been an avid reader. As a child, already I read a lot. In the 1960s, I bought a course in speed reading, inspired by US president John F. Kennedy, who had successfully used the system. The method consisted of learning to focus your gaze on one or several lines at a time instead of reading the text word by word. This sharpens your concentration and increases your reading speed. The method probably does not suit everyone, but it worked for me. Carola claims she remembers what she has read better if she reads slowly. For me, it is the other way around.

I followed the professional literature by making copies of articles and piling them up. For a long time, I read Current Contents, which appeared once a week and contained the tables of contents of all the important journals in biology and medicine. I marked the articles I wanted to read, and Hilkka made copies of them. I tried to read articles that directly concerned my own research as soon as possible. When it was time for our month of summer holiday in Finland, I brought the papers with me that I had not yet read. It was important for me to read many articles at the same time, and I always included articles from fields other than my own. I distributed to other members of the group the papers that were directly relevant to our research. They were often marked with blood stains from the mosquitoes I had killed on Sandö Island in Ostrobothnia. In the end, I went too far with my passion for reading research articles: I brought a suitcase full of reprints with me on our vacation. My family thought I had gone mad, and eventually I came to my senses.

Reading through a huge pile of papers enhanced my ability to bridge fields and associate. This practice also gave me a wider view of cell biology. New ideas most often arise by association. However, the capacity to associate requires practice to combine observations and data in new ways. And you need a broad spectrum of knowledge that can be activated when needed. When I am listening to an exciting lecture from another field, for example, there may suddenly be a flash in my brain, and a new idea is born.

During another vacation, I read two cell biology textbooks, one after the other. First, I embarked on the classic 'The Cell in

Development and Inheritance', written by E. B. Wilson in 1896. It summarised the first golden century of cell biology. I was impressed that so much had been discovered with the aid of only light microscopy and a few staining methods. It was already clear then that cells in all organisms are organised on the same principles. The data of the cell biologists provided convincing evidence of Darwin's theory of evolution. I was surprised that such an old text could still be so captivating. The second textbook had just been published, 'The Molecular Biology of the Cell'. It was a *tour de force*, extremely well written and, above all, well illustrated. It became the standard textbook all over the world, and our work was included in it! Here was the life cycle of the Semliki Forest virus in words and pictures. We were very proud. Every researcher dreams of getting their results into the leading textbook in the field. It was the last time I read a textbook from beginning to end, but this one was really worth the effort.

Biochemist Bruce Alberts, one of the authors of 'The Molecular Biology of the Cell', later came to EMBL to give a talk about centrosomes – the structures that are needed for cell division – which were the subject of his research. During my introduction to his talk, I told him that I had read his new textbook and found it very exciting. I noted that it documented how the last eighty years had revolutionised cell biology, making it a nice companion to Wilson's book. Still, I pointed out that Alberts' own speciality, the centrosomes, had actually been described better by Wilson in 1896 than in his own textbook, and he admitted that was true. Today, we know much more about centrosomes. This new knowledge has been incorporated into later editions of the book and has far surpassed the knowledge we had 1896.

Why do so many researchers lack interest in developments in fields other than their own? The virologists at that time were a typical example. They focused entirely on their own viruses and did not bother even with the closest relatives of those viruses. Semliki Forest virus belongs to the alpha virus family, and its nearest cousin is Sindbis virus. American virologists preferred to work with Sindbis and completely neglected our results with the Semliki Forest virus. For them, it was enough to be published in American virology journals, and there, you could get away with only citing research done in the USA. Strange but true. One might think that researchers would stay curious throughout life, but often this is not the case. We tried to keep abreast of research on all membrane viruses and could not understand that others could survive without a broader scope. We did not call ourselves virologists but cell biologists, and we followed the literature not only on enveloped viruses but also on cell organisation and membranes. This gave us a considerable head start and promoted our research.

Now that the Cell Biology programme at EMBL was delving deeper into cell biology, I thought we should recruit someone who knew how 'real' cell biologists worked and thought. At that time, in the Cell Biology programme, we were mainly membrane biochemists. None of us had a classical training in cell biology using morphology as the main approach. We managed to hire Kathryn Howell from George Palade's lab, which had trained many of the cell biologists of the past. Kathryn was a great addition to our programme and could keep us on track when we were straying too far beyond the borders of cell biology. She was an expert in cell fractionation and isolation of organelles, which, until then, had been applied only to tissues. We needed to find new methods to isolate cell organelles from cultured cells for biochemical analysis.

***

Our research with MDCK cells was progressing. We discovered that membrane proteins of viruses are sorted in the Golgi complex; more precisely, in a membrane network that we named the trans-Golgi network (TGN). Here, the proteins are sorted into separate containers for delivery to different parts of the cell. For example, the influenza virus membrane protein is sent to the apical pole of the plasma membrane, whereas the VSV proteins are delivered to the basal pole. We imagined that the TGN worked as a special distribution centre within the Golgi complex and that it handled not only virus proteins but also cell proteins destined for the plasma membrane. This proved to be the case. What's more, Gerrit van Meer demonstrated that the TGN probably sorted lipids as well. Now, we could move to the next stage of our research to zero in on the mechanisms by which membrane proteins and lipids are sorted in the TGN of epithelial cells.

At that time, a problem arose in the laboratory. Karl Matlin got into a row with another postdoc, Steve Fuller. It was a heated affair. The other members of the group thought that Steve was so troublesome he ought to be fired, but I disagreed. Steve was a physicist by training, intelligent but arrogant. In his first presentation at our group meeting, he displayed a set of mathematical equations for how protein sorting in the MDCK cell might work. We understood nothing, but that was the purpose. Steve confused us just to show how brilliant he was.

I spent a great deal of time making Steve understand that he, like all the others, was dependent on teamwork. He should learn how to work together with others and how to collaborate, and I promised to help him as his mentor. As a group leader, one must invest time and effort to smooth over conflicts between individuals in the group, lest the atmosphere of the whole lab deteriorate. If a conflict starts to dominate, everyone is affected. Cooperation is a skill that must be learned, needing both a trainer and a manager. In this case, I had to fulfil both roles. I managed to cool down the emotions of our two combatants so that peace was restored to the lab.

I also spent time with the group leaders in the Cell Biology programme to emphasise the importance of cooperation and to fuel our synergistic efforts. It was a constant balancing act. Too much advice was construed as preaching and was not appreciated at all. I tried to learn how to interpret the body language of my interlocutors to understand how they reacted to my attempts to persuade. I adopted the habit of interrupting myself to make a fresh start when my reading of the body language indicated that I was moving to the wrong wavelength. That earned me the nickname 'Kai with the unfinished sentences'. One group leader gave me Schubert's unfinished symphony as a farewell gift, which aroused general merriment and showed that everyone knew my peculiarities. With age, I gained more authority and got rid of this habit.

I spoke four languages every day. At home, we spoke Swedish. Carola and I encouraged Katja, Mikael, and Matias to speak Swedish among themselves, and they did. In many other foreign families at EMBL, the children soon began to speak German together, refusing to use their parents' language. I almost always had a Finnish postdoc with whom I spoke Finnish, as I did with Hilkka. English was the principal language spoken at EMBL and, with time, I also learned German. I had the feeling I hadn't mastered any language. Unceasingly, I tried to convince myself that what was most important was to be able to express what I wanted in a comprehensible way. Communication was more important than linguistic brilliance. But how painful it was to be confronted by TV cameras and be forced to improvise in English or German, to compare yourself with those who spoke English perfectly, like the Brits.

English remains my professional language, and I am happy that it has this privileged status, which facilitates communication all over the world. At that time in Europe, in fact, broken English was generally the common language among scientists. EMBL was funded by ten European member states, and I considered it my duty to spread the message about what we were doing by accepting seminar invitations. I wanted to demonstrate that their contribution to the EMBL budget was well invested and hoped that they would appreciate our role as a training centre providing services to the European community. Also, in these travels, I was a talent scout, looking for new collaborators and recruits wherever I went.

**

By keeping in contact with researchers in the EMBL member states, I also experienced how the working conditions of European researchers differed from country to country. Not only in Finland were resources scarce. I was surprised to learn that the Italians were in an equally bad situation. Once, I gave a lecture in Pisa at the elite institution *Scuola Normale Superiore*, which was founded by Napoleon in 1810. The buildings were beautiful, monumental, and steeped in tradition. The lecture hall was located in a palace, but the screen was pitifully small and the slide projector was so old that it hardly gave any light. It is not so simple to combine old and new. Italian biology was funded by the provinces. It had no central research council like those of other countries (the Academy of Finland, for example, which could distribute its funding, even if modest, to the most competent researchers in the nation). I find it difficult to understand this neglect of Italian science policy. Italian research funding is still minimal, and the result is that many successful Italian researchers go abroad to work. There are areas in biology and medicine where Italy shines, nevertheless, but considering the size of the country, more should be possible.

Telethon, a private foundation, invited me to join their scientific advisory board in 1999. They raised money to support biomedical research in Italy. The project was initiated by Susanna Agnelli who had been Minister for Foreign Affairs and was an influential person from an important family. The purpose was to fund research evaluated by experts while operating outside the corrupt Italian system. The first meeting of our expert committee, composed of four foreigners and six

Italians, did not bode well, however. Our task was to review the applications and select the best projects. As we considered the fourth project, the Italians joined forces and started debating in Italian. After their caucus, they proposed that the project be funded. We foreigners protested and threatened to leave the committee. We would only accept discussions in which the entire committee could participate and request transparency; otherwise, our contribution would be worthless. The chairman supported us and would not allow any more horse trading. In my experience, European and international committees are more objective than national ones.

Not only did poor funding impede research in Europe, but scientific inbreeding was another European problem. Gerrit van Meer came to my lab from the University of Utrecht, which had long been an international centre of lipid research. This research was led by Laurens van Deenen, a charismatic and inspiring personality. Regrettably, van Deenen ensured that his professors came from his own school, which gradually led to a situation where the whole centre was occupied by little van Deenens. This could not work in the long run. Soon the vitality of the centre was lost, and decline set in.

I saw the same thing at Christian de Duve's institute at the University of Leuven in Belgium, which was founded about the same time as EMBL. De Duve was a renowned scientist, a professor at Rockefeller, and a Nobel laureate. His new research institute studied cell pathology. The field was well chosen, and the institute had every chance of becoming a success, but this did not happen. In the 1980s, I was a member of the scientific council of the institute and observed the sad development. Just like van Deenen, de Duve hired largely his own students. What's more, he tried to attract research money from industry. There is nothing wrong with that, in principle, but in Leuven, it led to catastrophe. The industrial projects were secret and precluded collaboration within the institute. The scientific council advised de

Duve to change these practices, without success. Later, we heard that de Duve had buried our critical reports in his desk drawer; they never saw the light of day. Research institutions that do not bring in fresh blood from the outside suffocate from the inside.

***

In the long run it became tiring to try to keep in contact with institutions around Europe and the USA. Most scientists travel too much. Why do we do that? The most important reason is to exchange information. By meeting colleagues at conferences and at their institutions, we stay informed about the latest developments. Another reason may be to defend our territory rather than leave room for our competitors to move in. Often, we accept invitations in order not to be regarded as arrogant. I know of only one famous investigator who never travelled anywhere: Fred Sanger, winner of two Nobel Prizes. After retiring at sixty-five, he tended his garden and built boats until he was ninety-five. Maybe we will come to our senses after the COVID-19 pandemic and continue to use online seminars and workshops. We will always need conferences where people are together in one place, but a mix of online and in-person meetings would be healthier and more sustainable than the excessive travelling that I have experienced.

The invitations I received to give seminars and talks led to some unforgettable trips. Neither Carola nor I will forget the centenary of the Pasteur Institute in Paris. It was an incredibly lavish celebration sponsored by Air France. The festivities showed how important the Pasteur was as an institution in France. The celebration was opened by President François Mitterrand in the old Pasteur library with the renowned Finnish painter Albert Edelfelt's portrait of Pasteur hanging on the wall. Pasteur's reverently preserved apartment was in the same building. There were three oil paintings by Pasteur himself, which he painted at the age of thirteen. The paintings were so good that he might have become an artist, but he chose to become a scientist instead.

What in those times was called the 'Ladies' Programme' was exquisite. When the ladies visited the fashion house Nina Ricci, I slipped in, too. We were treated to an intimate fashion show in a small room and were given sketchpads on which to order the creations that interested us. Not many orders were placed, however; the prices did not quite match our wallets. But it was a fascinating spectacle. The models floated around us in fantastic creations. The intimate show combined the force of imagination with a shade of eroticism and resonated powerfully with me.

My own talk at the anniversary symposium was only so-so. It was being simultaneously interpreted into French, but I spoke so fast that the interpreters couldn't quite keep up. When I paused, the interpreters went on for so long that the whole audience broke out in laughter. I started by talking about Finnish Edelfelt's portrait of Pasteur. On the international stage, I have always wanted to emphasise that I am a Finn. Here, at least, that point became clear.

On another occasion, Ari and I were invited to a cell biology conference in Nairobi. Carola and Majlen also came along. It was held at the International Laboratory for Research on Animal Diseases (ILRAD), where the well-known American cell biologist Don Fawcett had worked. In a textbook of histology, he had published electron micrographs of cells and organs that were truly inspiring. After the meeting, we were taken on a safari in the Masai Mara wildlife reserve with a charismatic guide called David Drummond. David had been involved in filming the movie 'Out of Africa' and knew everything about big game and savannahs. He recounted his adventures during the Mau Mau rebellion that led to the independence of Kenya. He was one of the twenty-two British secret service agents sent to infiltrate the Mau Mau guerrillas. To pass unnoticed, they dyed their skins black. It seems incredible that this would fool anybody. Some agents were exposed when they had children, revealing their true skin colour. David told us that he had played an important role in the negotiations between the British, Jomo Kenyatta (the Prime Minister of Kenya from 1963–1964), and the Mau Mau guerrillas. We didn't know whether he was making things up or telling the truth, but his stories kept us amused.

The most fantastic of David's stories were about cheetahs. He claimed that cheetahs use twelve words, of which he knew eleven. One day when we were driving around in a jeep, he did a little experiment as practice for a planned TV programme. We drove up to a group of five cheetahs. David got out of the car and cautioned us to be quiet and not to move even if a cheetah climbed up on the hood. We hardly dared to breathe as we observed David chatting with one of the cheetahs. The situation was unreal, and we were scared to death, but all went well. The cheetah let David understand that the family was just a little nervous, as two lions were roaming in the area!

The days in Masai Mara were exciting. Carola and I slept in a tent allegedly used by Robert Redford during filming of 'Out of Africa'. David warned us that if we went outside in the darkness, we must first make sure that there were no sleeping lions outside the tent. A campfire was burning all night to discourage herds of elephants from stomping across the camp. David was a phenomenon, but, finally, he went too far. We heard that later, he had an altercation with a cheetah and was severely wounded. Shortly after our adventure in Kenya, I was sitting in an aeroplane to Edinburgh. My neighbour turned out to be from Kenya, and he knew many stories about David. He was, indeed, a celebrity in the country.

During the meeting at ILRAD, I met Onesmo ole-MoiYoi, who had a stellar career in science and was one of the leaders at the institute. He was a Masai and had grown up on the savannah in Masai Mara. An

American millionaire had selected him for education in the USA. He did so incredibly well that he was accepted to study medicine at Harvard and became a professor of molecular biology. I can hardly imagine a more breath-taking leap into a new world than his.

***

In the 1980s, I went to Moscow, in the then-Soviet Union, with a small group of researchers. We had been invited to a meeting by the leading Soviet transplantation surgeon, Valery Shumakov. The guest of honour was Emil Bücherl, inventor of the Berlin Artificial Heart, which was intended for use as a transplant. The trip was organised by a company that sold laboratory equipment to the Soviet Union. At that time, the country bought entire laboratories, completely equipped with everything from pipettes to ultracentrifuges. The state bureaucracy simplified their work by ordering the whole package.

I don't know why I was invited, but I was enthusiastic about the invitation because I had a mad idea. In the spirit of promoting peaceful relations between Finland and the Soviet Union, I imagined that it would be possible to start a process that would result in the Soviet Union becoming a member of the European Molecular Biology Organization (EMBO), as a means of furthering researcher collaboration in Europe. If the Soviet Union were to join, the other member countries of the Council for Mutual Economic Assistance (COMECON) might be allowed to follow. EMBO was founded before EMBL with the aim of organizing scientific conferences and distributing scholarships for postdocs. The idea was to break the monopoly of the USA in attracting young European molecular biologists by inspiring and financing them to do their postdocs in a European country instead.

I had talked to the EMBO leadership, but they thought my idea was foolish and didn't believe my initiative would interest the Soviet Union at all. But I also talked to the CEO of the company that sponsored our journey, and he was enthusiastic. Since they sold laboratories to the Soviet Union, he had excellent contacts with the Soviet authorities. He thought it would be easy to put me in touch with the appropriate ministries. He kept his promise and accompanied me to my meetings. He was dressed in an overcoat with enormous pockets filled with presents. Before I had a chance to open my mouth, he had lined up the gifts on the table, which significantly lightened the atmosphere.

I shared details with the ministries about the activities of EMBO and how the programme had stimulated research in molecular biology in Europe. EMBO had succeeded in advancing postdoctoral exchange between the European countries, and many young researchers now spent this important phase of their training in Europe rather than in the USA. Researchers all over Europe would profit from the Soviet Union joining EMBO. I also told them that I had discussed the proposal with Finnish President Kekkonen and that he supported the initiative. I could see that this aroused attention. Admittedly, it was pure bluff, but I thought that the claim served a purpose and would not damage relations between Finland and the Soviet Union. After that, we discussed trivialities, and I hoped that the administrators present would take the initiative to a higher level. How unbelievably naive I was! I had been indoctrinated about Finland's role as a mediator between East and West and thought that little me could work wonders.

The conference started every morning with a vodka in Shumakov's office. The dinners and the lunches were extravagant. In accordance with Russian tradition, we raised a lot of toasts to friendship and success. We also made an excursion to Zagorsk (which has now restored its old name, Sergiyev Posad). The most important Russian monastery, Trinity Lavra of St. Sergius, is in that town; it is the spiritual centre of the Russian Orthodox Church.

In Zagorsk, too, we drank vodka and ate caviar. We who came from Germany were especially welcome since German churches had paid for the maintenance of the old churches and monasteries in Zagorsk, which did not receive the support they needed under communism. On our way back to Moscow, the bus made yet another stop for more vodka, caviar, and speeches, according to established protocol. Bücherl, the guest of honour, had to drink so many toasts that our normally so distinguished professor totally lost his temper. "*Ich kann nicht mehr!*" he shouted over and over again. It was almost frightening. The Russian scientists were completely bewildered and asked me what had happened. What had they done wrong? They just wanted to be friendly. Bücherl had not had the training of Finnish politicians and could not cope with all the vodka served in his honour.

Yuri Ovchinnikov, the Vice President of the Soviet Union's Academy of Sciences, was a membrane researcher just like me. We had met previously at several conferences, but when I came across him at conferences after my visit to Moscow, he avoided me. I understood that I had become a *persona non grata*. There was never any hint from the Finnish side that I had messed up in Moscow, but my initiative came to nothing. The Soviets were not interested in international contacts. Too many scientists would probably have defected if they had been allowed to travel to the West. Not even contact between socialist scientists were encouraged. I knew a Hungarian researcher who told me that he had once been invited to Moscow but had sat in his hotel room most of the time, as his attempts to meet Soviet scientists were blocked. A few researchers were allowed to travel to the West, but that privilege was restricted to the top nomenklatura. This I understood only in retrospect.

***

Leevi Kääriäinen, who during all these years had struggled in Finland, finally ran into trouble. The Academy of Finland would not extend his contract. I had to collect a petition from abroad, signed by David Baltimore, among others, to make the Academy change its decision. We pointed out that top-level research requires continuity. Researchers cannot be without a salary for a year or two and hope for better luck next time. The budget appropriations for research in Finland were too small to maintain decent funding for the fields competing at international level. The petition was effective, and Leevi was not obliged to move abroad.

The brain drain continued, however. When Ralf Pettersson, a member of the Kääriäinen–Renkonen–Simons troika's team at the Haartman Institute, was appointed director of a new Ludwig Institute within the Karolinska in Stockholm, there was uproar in Finland and a campaign in the press to bring about a change. Ralf pointed out that the Medical Research Council of Sweden distributed the equivalent of forty-one million euros annually compared with the paltry four million euros of the Medical Commission at the Academy of Finland. The pressure was mounting in Finland. Paavo Riekkinen, Professor of Neurology at the University of Kuopio, bragged about himself in the popular Finnish magazine *Suomen Kuvalehti*, however, suggesting that researchers who moved abroad had only themselves to blame. They were simply incapable of acquiring external funding for their research. In his words, "*Turha joukko vaeltaa*": the lightweights emigrate. In due course, Riekkinen received a prison sentence for embezzlement of some of his research grants for private use.

It turned out to be a wise decision to invest in five universities. It slowed the migration to the south of Finland, which would have polarised economic development and complicated the political situation in the country. In the far north, close to the polar circle, the University of Oulu was gaining prominence in biomedicine. Despite a tight budget, Kari Kivirikko had established a successful international group in connective-tissue research. Kivirikko, Karl Tryggvason, and

Reijo Vihko also developed the virtual Biocentre Oulu, where I was chair of their Scientific Advisory Board. It was a pleasure to see how enthusiastic researchers, young and old, succeeded in making Oulu a centre of medical research. How was that possible? The professors cooperated and supported one another!

In Helsinki, things were different. There, the scientists quarrelled, and I was pleased to have left that hornets' nest. Leevi was trying to establish an institute for gene research, but he met with stiff opposition. As usual, the conflict was about resources. Now that peripheral universities like Oulu were drawing more funding, resources became increasingly scarce for everyone. Not until the 1990s, when the government finally started investing seriously in research, did Finland move from this valley of misery to new heights. Antagonists were reconciled, and new institutes were founded. I compared the competition between universities to the Finnish Ice Hockey League. The upstart Oulu was the champion for some time, but then Helsinki rose again and secured its top position in the Biomedical League. Competition may have positive consequences, but cooperation is more effective. Finnish research had the wind in its sails thanks to its exemplary policies in the 1990s.

***

Our family spent all the summer holidays on the peninsula of Sandö on the west coast of Finland where Carola's family traditionally spent their summers. For German families, it was a recurring dilemma where to go for vacation. We had no such problem. We stayed in an old granary, which we had moved to Sandö. The logs were from 1714. It was cramped but cosy. We liked that it was so primitive. There was no electricity, and the water came from a well. Carola's mother and siblings also had their summer cottages on Sandö, so it was a little village by the sea. Beautiful nature, clean air, good company, and parties – what more can one wish for? Even the clouds were different from those in Heidelberg. In Finland, the clouds were rich in contrasting colours and formed fantastic three-dimensional constellations against the blue sky. I always enjoyed watching these beautiful clouds. When Finns were named the happiest people in the world, I was a little bewildered at first, but it's not surprising considering how much marvellous and untouched nature is available to every Finn.

The whole family went skiing in the Alps for our winter holidays. We were out on the slopes in all weathers, even in the freezing cold and wind. On one such bitter day at the Swiss resort Flims, we took a break at an alpine hostel where we met a group from Heidelberg. They had all stayed indoors because of the cold. One of them was Wolf Forssmann, with whom we used to play fathers-and-sons football. He was a go-getter who always had to win. Once, by mistake, he gave me a black eye during a game. Forssmann came along with us to ski but could not convince the others to join in. We went all the way to the top of the glacier, although there were such strong gusts of wind and snow that we barely managed to stay on the ski lift. When we were at the top, Wolf wanted to prove his mettle by skiing straight downhill across the glacier. Typical Wolf.

Wolf took after his father, who had performed the first heart catheterization in the world – on himself, assisted by a nurse. His boss had regarded the attempt as too dangerous and had forbidden it, but Werner Forssmann was not to be stopped. When he published his feat, he was working as an assistant at the Charité – Berlin University of Medicine with the famous surgeon Ferdinand Sauerbruch. Sauerbruch thought that the ground-breaking work was a pure circus trick and dismissed him. It took some time before Forssmann was recognised for his pioneering achievement, but, at last, in 1956, he was awarded the Nobel Prize. Boldness is an important quality in a researcher. You need pragmatic self-confidence to overcome the obstacles that confront you. Who dares wins!

***

For me, it was important to broaden my background by reading not only about science but also about other areas. One important source of information for me was The New York Review of Books. This journal has been my faithful companion since the early 1970s. Many of the most prominent intellectuals in the USA published articles about literature, economics, politics, art, architecture, and science. The texts were long but well-written and a pleasure to read. If a new novel received an excellent review in the journal, I put it on my reading list.

Later, when The London Review of Books appeared, I subscribed to that as well. This review gave a British perspective and offered exciting intellectual entertainment. If a book was strongly recommended by both journals, this became the new criterion for my reading list. The people around me scoffed at my habit of referring to books I had not read and accused me of only reading reviews. I admit that was often the case. I certainly had no time to tick off my whole reading list. Living in Germany naturally meant that Carola and I also kept up with German politics and cultural life. Our favourite newspaper was *Die Zeit*, which published the best reviews I could find of German books.

Even from the viewpoint of my own research, it seemed important to broaden my knowledge base. It increased my self-esteem. I feel respect and gratitude to all those who give so much of themselves to make this world more understandable and more enjoyable. I know many successful scientists who are generally well-informed. Of course, other researchers do well while remaining narrower in their focus, and some are even obsessively myopic in their interests, but this would not work for me. I feel there is a trend towards less general reading.

Scientists are often too busy with themselves. I feel that there is something missing when I do not read beyond my research interests.

***

My father died in December 1986. He had been healthy all his life, but only three months earlier, he was diagnosed with an inoperable stomach cancer. We were close; ever since childhood, I had talked and discussed everything between heaven and Earth with him. Often, we stopped only when my mother Rut interrupted our palavers and told us to go to bed; she thought we needed sleep as much as talk.

I was at a meeting in Sorrento, Italy, when I got the message that my father's condition had deteriorated. I went down to the market and bought the finest delicacies I could find. Then I flew directly home to Helsinki. Sadly, it was Lennart's last meal. After that, he had to be fed through a tube at the Surgical University Clinic. Despite everything, his final days were unusually harmonious. He chatted with everyone who came to visit him and made the farewell meetings memorable. The whole family went to Helsinki, so the children, too, could be together with their grandfather for a few days. They had often stayed with him during the summers. I can only hope that one day, I shall be as strong as he was and manage to follow his example.

Carola and I tried to keep in contact with friends and relatives in Finland as well as we could. It was our native country, and we did not want to lose these ties. I stayed in touch with my schoolmates from Ågeli, mainly Ralf Nordgren. Ralf's parents were active communists. In our school days, we argued constantly about communism. I saw no future for the Soviet system, but Ralf was unable to liberate himself from Stalin and the dictatorship of the proletariat. Our wrangling sharpened my interest in society and politics. Ralf became a writer and had some success, but his life was not easy. His whole family had problems with alcohol, and that was Ralf's fate as well. We rarely wrote letters, but when, recently, I browsed through old notebooks and papers, I found one that I want to cite here.

Ralf had written the letter after a meeting at Restaurant Elite in Helsinki where, as usual, we had had intense discussions. I had blurted out that it is not as important to ask why as it is to ask how. Ralf lived in an old house beside a small river outside Björneborg on the west coast, where he had discussed with his neighbours whether why or how is the more important question. Most of the practically minded neighbours, who were farmers, supported how. Ralf had also written to the philosopher Esa Saarinen, who had voted for him. In the world of the arts and humanities, how may not be so important. In this letter, Ralf wanted to know why it was so important for me. "So here I'm sitting wondering what you really meant, and if my recollection of your proposition is incomplete, please correct me. But anyway, I think it would be interesting and important to know why (!) you insisted so intensely on how questions."

Scientists look for important and interesting problems that can be solved. We are problem solvers. We do not ask why the problem exists; we ask how we can solve it. This attitude has propelled the Western world forward ever since the Age of Enlightenment. When the church lost its power in the seventeenth century, the floodgates opened for innovations that have made our world unrecognizable, in good ways as well as bad. As a species, *Homo sapiens* is expert in answering how questions. When this potential was liberated, the rate of discovery and invention accelerated. It is illuminating to consider how average life expectancy has changed. For several centuries, it was little more than thirty years. It began to increase in the nineteenth century, and today, on average globally, it is seventy-two years!

Why questions cannot be solved in the same way? Often, in fact, they are unsolvable. Take, for example, the question of why we are here. When children ask "Why, why?" their parents are often irritated because we cannot answer their questions. Why do children ask so much? Because they are curious. Why, then, are we curious? The why questions grind on infinitely, whereas how questions are formulated so that they can be answered.

Still, we must understand that the answers to questions provided by science and research are not sufficient to satisfy our needs. A full life requires more than that. Here, our culture often fails. Western civilization gives few satisfactory answers to the emotional problems that we must come to grips with individually and collectively. I have no copy of my reply to Ralf and do not remember what I wrote. Now it is too late; Ralf has passed away.

I tried to follow the cultural debate in Finland. In 1986, Finnish philosopher Georg Henrik von Wright published his essay collection 'Science and Reason'. He was one of my favourite writers. After the debate about the two cultures in *Hufvudstadsbladet*, von Wright invited me to his home one evening. It was exciting to discuss with Finland's most famous philosopher. He had worked with Ludwig Wittgenstein and became his successor at Cambridge, but he returned to his professorship in Helsinki in 1951. His essays were easy to read and not at all heavy in the way philosophical texts are usually.

'Science and Reason' is a critical analysis of how science, through technology and the industrial forms of production, has affected our lives. Von Wright emphasised that our world had moved in the wrong direction. He realised early on that the destruction of the environment might become fatal and that our unbridled consumption threatened our entire civilization. Then, as now, the threat of total annihilation by nuclear war was something no one could neglect. Von Wright predicted that *H. sapiens* would perish and disappear, as has been the fate of so many other species before us on this planet. His book has a pessimistic, almost frightening and gloomy tone. Despite the gloominess books of this type induce, they are important to make us reflect on where we are heading.

In 1993, von Wright published a new book, 'The Myth of Progress'. An article in the Swedish newspaper *Svenska Dagbladet* under the headline 'Beyond all reason' criticised severely von Wright's pessimism in this new book. Politicians like Carl Bildt, the Swedish premier, could not accept von Wright's prophecy that the flaws of our civilisation would to lead to the extinction of humankind as a species. Bildt pointed out that such predictions are dangerous. They lead to resignation and despondency. Society could sink into apathy or protest against novel technological developments that were badly needed.

For me personally, progress was no myth. In one hundred years, my family's living conditions had improved radically. My maternal grandfather Abraham was one of twenty children by the same mother in Petalax in Ostrobothnia. Most of the siblings died, but Abraham managed to emigrate to the USA. There, he became a Baptist preacher and somehow accumulated enough money to return to Vasa, where he founded a shoe shop and a shoe factory.

My paternal grandfather Mikael had only three siblings, but that was because his father, my great-grandfather, had died young. During the famine years of 1867–1868, my great-grandfather drove by horse and carriage from Vörå in Ostrobothnia to Kristinestad, more than 100 kilometres, to buy cereals. In Kristinestad, a shipment of cereals from Russia was expected. The ship never reached its destination, and great-grandfather got into a brawl after drinking in the pub. He had to spend a week in prison on bread and water because he had no money left to pay his fines. When he came home, he ate, probably had a bowel obstruction, and died. Grandfather Mikael and his siblings were placed in different households in the village. Poor children had to work for their living, and they walked in shoes made of birch bark because they could not afford leather.

Still, Mikael was fortunate in his misfortune. The farmer he was working for was a kind soul and gave him enough money to buy a ticket to the USA. Mikael became a miner in Utah and managed to save enough money to go back home to Vörå and buy a small farm. This was every emigrant's dream in Ostrobothnia, but few succeeded. When my father Lennart retired, he became interested in genealogy. Despite exhaustive searching, he found not a single academically educated person among his ancestors. Since time immemorial, we had been peasants and farmers.

Von Wright overemphasised the negative consequences of technological advances. Today, most of the world's population live healthier and more prosperous lives than ever before. Most can read and write, and global life expectancy has increased steeply since the end of the nineteenth century. Nearly all have seen improvements in their material living conditions, although inequality is on the rise and threatens the stability of our societies.

Yet von Wright was right in that we are on a path that has taken us dangerously close to the edge of a precipice. Every technological advance has unintended negative consequences. We see these today. Our society is no longer sustainable. The use of fossil energy heats our planet and will make it uninhabitable for humans if we do not change our way of life. Our only way out of this dilemma is to renew the technological basis of civilization. This is a Sisyphean task, but one that *H. sapiens* must accept. We cannot turn the clock back and hope things improve by themselves. We have condemned ourselves to eternal innovation.

The fall of the socialist system illustrates this thesis. The bankruptcy of the communist regimes was caused by their inability to renew their technology. Their societies became static. In contrast to von Wright, I think that we are compelled to accept the damnation of eternal innovation. Actually, it is not a damnation, but an exciting challenge for those who are creative and productive. *H. sapiens* is a curious animal. Many of us could not endure living in paradise. We need stimuli and challenges to find meaning in our lives. This quality makes us unique but also dangerous to our environment. All other species of the genus *Homo* finally vanished; only *H. sapiens* survived.

Our curiosity helps us to solve the most intractable problems but can also drive us to disaster. Von Wright writes "One day humans will certainly cease to exist as a species, if it happens after a hundred thousand years or a couple of centuries, is a trifle in the cosmic perspective." He was surely right. Despite his prophecy, our task is to continue to find solutions, and not only technological solutions. We need positive impulses from the arts and humanities to help us reform our lifestyle and change our attitude to humankind and to life on Earth.

Myself, I remain an incurable optimist. Our curiosity, our instinct for solving problems, and our innate cooperative abilities will save us, all doomsday prophecies notwithstanding. This is my way of looking at life.

***

At EMBL, life went on. I could glimpse a breakthrough in our attempts to discover a mechanism for how the TGN membrane is remodelled to direct the molecular traffic to different destinations in the cell. An answer to this question was crystallizing in my mind. The exciting element of the concept was that both lipids and proteins might be involved. This was, of course, what I had hoped for because we included both in our analysis.

The novel idea was that glycolipids (lipids covalently attached to a carbohydrate molecule) interact with proteins during the sorting process in the TGN. Vague clues from the literature suggested that glycolipids associate with each other; why not then with apical proteins as well? The basic sorting principle, I hypothesised, was that glycolipids gel together with proteins destined for the apical membrane, forming nanodomains in the TGN membrane. These nanodomains bulge out to form membrane vesicles. The vesicles are then transported to the apical plasma membrane of the MDCK cells to build up the apical pole, where glycolipids are known to be enriched.

But this was no more than a starting point; we needed data. I felt that my hypothesis opened totally new perspectives on membrane function. I had an incredible feeling that I'd hit on something really novel. It seems trivial to me when I write about it now, but at the time, it felt great. We had to proceed.

Lennart Philipson had successfully introduced the new structure of programmes at EMBL, so the organisation now ran more smoothly. He had also initiated a bioinformatics programme, which would become important for all of Europe. The DNA revolution had generated methods for sequencing the genomes of humans and any other species. Before the information could be used, however, the DNA sequences had to be deciphered and interpreted. During Philipson's time, the bioinformatics programme moved from Heidelberg to Cambridge and became an EMBL outstation. It was a great success. This type of project demonstrated the value of having a European institute. The first generation of bioinformaticians in Europe was educated at EMBL, just as the first molecular cell biologists were trained there during Kendrew's time.

Since most positions at EMBL were limited to no more than nine years — a new rule — there was a constant turnover of scientists and technical staff who could return to their home countries or other destinations. EMBL would thus provide for Europe staff trained in emerging areas of the molecular life sciences. This also meant that, as someone with a permanent position, I attended many farewell parties. Friends at EMBL regularly moved on to other places. Although many of us would still meet at conferences, seeing good friends leave always caused a tinge of melancholy. Researchers are nomads, which is tough on our families. When a group leader left the Cell Biology programme, we always had a big party. Gareth Griffiths adopted the role of master of ceremonies, and I participated in all the variety shows we performed, which helped to alleviate the gloom of the farewell. I never reached Gareth's level, but when we performed together some of his talent rubbed off on me.

Once, we were together at a conference in the USA where the elite of the membrane-trafficking field was gathered. Gareth gave a lecture and included several comical elements. Every third slide made the audience laugh. The talk was a sensation, the like of which I have never heard or seen since. The listeners roared with laughter. Some nearly fell off their chairs. Afterward, however, several researchers complained to me, "This is not OK, Kai. Naturally, one may start with some humour, this we all often do, but the entire lecture cannot be a long joke." Why not, I wondered. I saw how amused you were. A good laugh does no harm and may prolong life.

***

On 9 November 1989, the wall fell between the socialist GDR in the East and the democratic Federal Republic of Germany in the West. We watched the events on television, and soon all our neighbours gathered in our living room to celebrate the historical occasion. We were happy that everything had been so frictionless and agreed that we had to thank the Soviet leader Mikhail Gorbachev for that.

I was soon confronted with the fall of GDR socialism at close range. On behalf of the German Science Council, I spent four weeks in 1990–1991 with the commission evaluating the future of the East German Academy of Sciences in the field of biomedicine. In the GDR, as in all the socialist countries, most research was done in the Academies of Sciences, not at the universities. Six thousand researchers and employees were now thrown out onto the streets, and we were expected to decide their fate. It was an impossible and inhuman task. We travelled across the former GDR and saw how catastrophic the situation was. Practically all the old buildings were dilapidated. The *realsozialismus* philosophy of economic planning built mainly charmless grey residential blocks known as *platten*, and the countryside was not much better. The GDR produced insufficient building materials to renovate old houses and buildings, although leaking roofs were repaired. Despite this, the country was considered to be the most effective of the Eastern bloc countries that made up COMECON, and it occupied fourteenth place in the world economy. This did not reflect the real situation, however. When the system collapsed, the reality was revealed.

The commission was tasked with evaluating biomedical research at all academic institutes. We found that the level of research was way below that of West Germany and of western countries in general. Researchers had been isolated with no access to literature or equipment. There was no commercial infrastructure to supply them with the products required to do modern research, as we had in the West. With this lack of resources, there was no way that the researchers could keep up with us.

The same was true of industrial research. Fertilisers, pesticides, and herbicides for the entire Soviet bloc were produced in the East German towns of Bitterfeld and Leuna. The same chemical products, with the same pollutants had been produced since 1954 with the same production technology. For nearly half a century, these pollutants had been spread over the fields in the GDR and the rest of the COMECON block. What serious health problems had this caused the population? In fact, life expectancy started to stagnate in all COMECON countries during the 1970s and was about two-and-a-half years lower in the GDR than it was in the Federal Republic when the wall fell. The pollution of the air, water, and soil was alarming.

When we arrived in Leipzig, I stood on a downtown street corner one afternoon, looking at the people passing by during rush hour. I thought they looked unwell. The air was thick with brown-coal smoke, the food was unhealthy, and the health care was inadequate.

Microelectronics were another example of the backwardness of GDR technology. In 1989, before the fall of the wall, the first one-megabit microchip was presented with great ceremony in the East German city of Dresden with the East German leader Erich Honecker as guest of honour. The chip was produced manually and could not be mass produced like similar chips in the West. Innovation was conspicuously absent, and this was often compensated for by industrial espionage. It did not help much because the information obtained was rarely used effectively.

A modern society cannot function without unceasingly improving the existing technology and inventing new technology. A society that does not renew itself is condemned to bankruptcy in the long run. Think, for example, of antibiotics, one of our finest innovations, which almost eradicated the problem of bacterial infections all over the world – until bacteria took their revenge by developing resistance to antibiotics. The drug industry neglected this problem, but newly founded start-ups are now working intensely to find solutions. This dynamic is one of the drivers of the capitalist economy and produces innovations that often correct the negative side-effects of technology. It may sound mad, but it works. The socialist dictatorships were unable to sustain this dual-edged dynamic.

Our commission tried to come up with ideas for how the research structures of the old GDR might be modernised. We saw it as an opportunity to introduce reforms that would benefit the West German system as well. One useful reform would have been to move some research activities from the Academy institutes to the universities. West Germany also maintains too many research institutes outside the university system. These institutes suffer from the absence of students because research is stimulated by teaching and by the supply of young talent. Many university teachers might disagree, but having elite research alongside teaching is advantageous if the teaching loads are not piled too high.

Our proposals did not meet with approval. The government of Helmut Kohl accepted no changes beyond those written down in the treaty between the GDR and the Federal Republic. In practice, the West German system was implanted in its entirety in the East. This was brutal and still disturbs the harmony between the two parts of the country. With a more open policy, the wave of enthusiasm that arose when the wall fell might have worked wonders. If West Germany had understood what an opportunity this was, new initiatives and experiments might have stimulated the economy and led to increased cohesion and innovation. This would have benefitted both sides.

***

Under Philipson's leadership in the late 1980s, EMBL introduced a programme for doctoral training modelled on the American system. Doctoral training in Europe was not in good shape and needed reform. Professors and docents had the right to accept the students they wanted and were free to arrange their mentoring and supervision as they wished. Usually, it worked, but if problems arose, the student had no alternative but to adapt or leave. American universities took doctoral training more seriously and had graduate programmes that guaranteed a certain quality. If problems arose within their programmes, the student was not left to cope with them alone.

It took some time before we managed to make the EMBL programme attractive. Most doctoral students in Europe still did their thesis work where they had begun their studies. We had to give lectures and briefings all over Europe to attract talented candidates to our programme. Once, I went to Athens for such an event together with

Iain Mattaj, the coordinator of the Gene Expression programme. We were upgraded to first class on the flight along with the rock group the Scorpions, who were giving a concert in Athens. Iain and I happened to be the first out of the plane and were met by a big crowd of screaming fans, who took us for members of the band.

Our reception at the scientific event was somewhat different from that of the Scorpions at their concert, but we did manage to interest a few candidates in the EMBL PhD programme. Soon, it became a formidable magnet attracting superb doctoral students, who strengthened the laboratory's research. Today, hundreds of former EMBL doctoral students are prominent researchers all over the world.

***

I was struggling with my research. We were trying to identify the protein machinery in MDCK cells that sorts virus proteins to the polarised plasma membranes of the cell. Hilkka and I started to isolate the membrane structures that formed in the Golgi complex, in search of proteins responsible for sorting the correct cargo proteins and lipids to the apical plasma membrane. We had devised a method of opening the cells so that they spat out the membranes we needed. I suspected, however, that the approach we had come up with was going to fail. We would probably not be able to isolate sufficient membranes to identify the apical sorting machinery.

Membrane research is difficult. Our biggest problem was that the methods available were not what we needed. In my notebooks, I see recurring expressions of despair that everything was moving so slowly. The comments always end with my battle cry "Struggle on!", which sounds better in my mother tongue; I usually wrote my notes in Swedish. I tacitly add that nothing advances by itself.

I found some consolation in the essays written by the Hungarian–Swedish cancer investigator Georg Klein in Stockholm. In his portrait of chemist Louis Pasteur, Klein describes how Pasteur spent every afternoon walking around the library in his high quarters, mumbling something to himself. His collaborators kept wondering what he was saying. One day, one of them managed to pick up what he was repeating over and over again: "*Il faut travailler, il faut travailler.*" ("You must work.") When I first read it, I was confounded by Pasteur's mantra. Today, I understand him more than well.

By working myself on the apical machinery project, I hoped that I could somehow encourage the rest of my group to engage with this problem. And this time, it worked! The lab came up with new methods for characterizing the apical sorting machinery. We used two-dimensional gel electrophoresis to separate and identify the proteins from the membrane structures that we isolated. In the end, we also cloned the DNA encoding the proteins we found. It was a slow process, but we gained ground. Temo Kurzchalia, one of the postdocs involved, introduced a new ritual: every time he carried out an experiment that he regarded as important, he put on a tie. The ritual spread throughout Cell Biology. Seeing the young researchers, female as well as male, dressed in ties stimulated us all. We took it as a symbol of the unbearable lightness of research, to paraphrase Milan Kundera; bright spots in the daily chore of doing research.

Our up-and-coming research field, membrane trafficking, was attracting talent. One of them was Jim Rothman, later a Nobel laureate. He was always on the go and critical of the work of others. However, he often turned a blind eye to problems in his own research and presented his results in the most shining light. When he presented his work to non-experts, he made it sound as if all the questions had been answered. Researchers working in other fields of molecular biology who heard Jim's seminars wondered why we were wasting time on questions that had already been solved by Rothman.

To counteract this annoying situation, Ira Mellman at Yale and I decided to write a review of the field. We warned Jim that if he did not want to discuss his own data, we would bring forward the criticisms that were circulating in the field. This had no effect on Jim. He went on in the same style.

As promised, Ira and I wrote our review and sent the draft to other researchers in the field, Jim among them, for comments. Many gave us valuable feedback, which we incorporated into the manuscript. Jim's feedback, however, was longer than our entire manuscript. We had deliberately sent out a hypercritical version of the review to show that, notwithstanding brilliant Jim, there were still important unresolved questions. The final text was more balanced, of course. Our title was 'The Golgi complex, *in vitro veritas*?' *In vitro* refers to cell-free experiments 'in glass' test tubes, and the title alluded to the fact that Rothman was a pioneer of these methods.

***

By then, Cell Biology at EMBL had fully adopted recombinant DNA technology in its research repertoire. One postdoc worked with a genetically modified variant of the vaccinia virus, which is a close relative of the smallpox virus. By mistake, she pricked her finger with a needle that contained the recombinant virus. To check whether everything was okay, she went to the university clinic. This happened in the evening hours. When the doctor on call heard that a genetically manipulated virus was involved, panic broke out. They raised the alarm and wanted to put her in quarantine. We knew that the virus was harmless and we had this confirmed by the Centers for Disease Control and Prevention in Atlanta, USA.

This did not end the crisis, however. A political dimension emerged. The head of security at EMBL, responsible for DNA research and biocontainment, had recently been dismissed, and I knew that he was a good friend of a member of the state parliament of

Baden-Württemberg representing the Green Party. The Greens fought hard against all kinds of gene technology. We scientists assumed that recombinant DNA work at EMBL, as a European institution, was not subject to all the paragraphs of German law, so we did not bother with every bureaucratic detail, only with those that belonged to good laboratory practice. However, it turned out that our Director General had informed the authorities that EMBL followed all German regulations regarding recombinant DNA work, and our former head of security knew this. Now we were in trouble. As head of Cell Biology, I was personally responsible. I realised that I needed help. If this mishap became public knowledge, we would all be in the media.

The head of personnel at EMBL, Konrad Müller, was an unusual administrator. One might say he was a 'fixer' in a positive sense. He had political experience from both NATO headquarters and Bonn, the West German capital. In Bonn, he had worked for Herbert Wehner, the legendary whip of the Social Democrat parliament group who had great political influence. Yet Konrad decided to leave the corridors of power to return to his hometown of Heidelberg. What was most amazing about Konrad was that he knew so many people in important positions. Konrad found a way to prevent the Greens from taking the EMBL DNA security issue to the state parliament. How he did this, I don't know, but what a relief! After this scare, EMBL meticulously implemented all regulations and reported all experiments to the regional security commission for recombinant DNA methodology, as required.

In 1992, Fotis Kafatos succeeded Lennart Philipson as Director General. Kafatos, a Greek, had had a brilliant research career at Harvard. When we met for the first time at a conference in the USA, he was dressed in a slick black satin suit and made a great impression on me. He was a pioneer in DNA technology and a founder of the field of insect molecular biology. Now, he was studying malaria.

Kafatos shuttled frequently between Crete and Harvard. On Crete, he had been involved in building a new university and had devoted his energy to a new biological research institute that he founded. Whereas Philipson had not been interested in Brussels or the European Union, Kafatos, by contrast, was well connected there. At EMBL, we looked forward to the prospect of strengthening our EU contacts with Kafatos as our chief. Fotis and I hit it off immediately, and I hoped that I could help him to prepare for the future of EMBL.

***

When I was invited to a workshop in 1993 by Biogen in Geneva, I had no previous contact with the industry. Biogen was one of the world's first biotech companies. The workshop was arranged by Ken Murray, who had worked at EMBL but was now a professor in Edinburgh, UK. He had been one of the founders of the company in 1978 and was a member of its board. Biogen commenced its activities in Geneva but had since moved to Boston, USA, because the concept of start-ups was new at the time, and the conditions for establishing such companies were difficult in Europe. Start-ups depend on venture capital, which was hard to find in Europe; capital was invested preferentially in the USA, where the start-up culture was in full swing. Another problem was that Biogen had not succeeded in establishing contacts with the pharmaceutical giants in Europe, which at that time did not negotiate with dwarfs like Biogen. Therefore, Biogen moved to Boston, where it was much easier to build up its operations.

When I began my career in research, biology was largely free from commercial interests. Our communities had little contact with industry. So, I was surprised right after the Geneva meeting to receive an invitation to become a member of Biogen's scientific advisory board. If I accepted, I would even receive payment for this activity. The offer gave me a headache. Should I really work for the industry for money? As a child of the student revolt of the 1960s, I felt reluctant, but at the same time, I wanted to learn more about how the achievements of molecular biology could be put into practice. We do not work in a vacuum.

After long discussions with Carola and Konrad Müller, I decided to accept the position, and I never regretted it. The advisory board met four times annually with Biogen's management team to discuss the company's established and new projects. Biogen also invited various external start-up companies to present their research in order to see whether they might be interesting to them. It was fascinating to see how they presented their products. After the presentations, the board analysed the scientific background of the projects to assess the quality of the research and development, and the management discussed the economic situation of the start-ups. All the start-ups that presented to us came from the USA. They tried to commercialise the best ideas generated by molecular biological research. Nevertheless, most projects were fraught with so many weaknesses that Biogen did not want to engage with them. There was simply no standard recipe for successful commercial projects in biomedicine. Everyone had to feel their way forward. The message I took home was that even if Europe was way behind, we could still be successful. Biotechnology was in its infancy, and there was plenty of room for new ideas.

For some time, Biogen had been on the brink of bankruptcy. CEO Walter Gilbert, a renowned molecular biologist, had not succeeded in developing a drug candidate that could propel Biogen into profitability. Then, Jim Vincent, former Vice President of Abbott, a leading healthcare enterprise, took the helm. When I asked why he took on a near-bankrupt company, Vincent answered that a company founded by such prominent molecular biologists simply must have hidden treasures in its portfolio. Soon, he led Biogen to a top position among the new biotechnology companies by buying back the rights to interferon-beta, which Gilbert had disposed of. Biogen developed interferon-beta (brand name Avonex) into the first drug against multiple sclerosis and established itself as a global company.

Alick Bearn, who had been my postdoc supervisor, was also a member of the Biogen board. He had left his professorship at Rockefeller to become Chief of the Clinic of Internal Medicine at Cornell University in New York and, later, Vice President of Merck Sharp and Dohme, where he oversaw clinical studies. Alick must have played some part when I was accepted into such an illustrious business circle. Maybe he had even told the story of our party in New York, where he was exposed to Lala Eriksson's grand socialist speech. Biogen CEO Jim Vincent was as jovial as he was conservative. He jokingly called me 'our communist', as in, "And what does our communist have to say today?" Despite this denigrating epithet, Vincent had accepted me.

One project that we discussed intensely was Biogen's plans for the commercial development of gene therapy. The scientific advisory board was chaired by Phil Sharp, who was known for his research on split genes. He was clearly in favour of the project, whereas I thought it was too early to enter such a complex field. Biogen invested millions into a centre for gene therapy in Philadelphia, USA, but the investment failed. The first patient treated there with gene therapy died. The pressures to move the project into clinical testing had been overwhelming. The protocols were not ready. After this failure, clinical trials for gene therapy worldwide had to take a long break until the research was so advanced that the prospects of success were all but assured. Today, hundreds of gene therapy trials are underway, and some treatments have been approved by the United States Food and Drug Administration.

***

It soon became clear that working under Fotis Kafatos was not so easy. He was an excellent chairman of meetings and host of delegations and visitors. He also successfully pursued the interests of EMBL among politicians and decision-makers. But developing EMBL research activities further was not his top priority. He often asked me for advice, but he seldom followed up. Cell Biology came up with a plan to establish a new programme in the Physics of Life. Eric Karsenti, one of our group leaders, was the driving force. He came from Paris, where the endeavour to combine physics with molecular biology was already well underway.

Many of the pioneers of molecular biology had been physicists before entering biology. Danish physicist Niels Bohr had urged young physicists to take on the basic questions of biology. He predicted that there were new fundamental laws to be discovered in biology. That turned out not to be the case; instead, the first molecular biologists discovered the chemical structure of our hereditary material, deciphered the genetic code, and found out how our genome works. This was revolutionary progress and started a new era of molecular biology that spread into all realms of biology and medicine.

We thought that biology was ready for a new revolution led by physicists. Eric Karsenti invited several interesting candidates to present themselves. Regrettably, our initiative found little resonance. Kafatos was not interested and granted no funds for the project. If EMBL had embarked on this venture, it would have invigorated the institute for years to come. We might have had success like that when Lennart Philipson launched bioinformatics, and EMBL's first generation of bioinformaticians led the field in Europe. EMBL's solid reputation would have helped us attract young talent from all of Europe to a physics programme. There, we might have been able to shape the fusion of physics with the life sciences once again. But EMBL missed that chance.

In 1992, I took part in a conference in Exeter called 'Europe into the Third Millennium', which discussed the European integration process. There were twenty-five participants, and I was the only scientist. The Icelandic President, Vigdís Finnbogadóttir, also participated in the meeting because Iceland was considering EU membership. The atmosphere was optimistic, and all participants believed in the future of the EU: soon, we thought, national borders would vanish, and we would have a common EU currency.

The EU was Europe's response to the war and misery of the calamitous twentieth century. We thought we could avoid similar catastrophes in the future by uniting the countries of the continent. But the EU faced so many challenges. How should we incorporate the eastern European countries that wanted to join? How should we prevent the destruction of the environment and the looting of Earth's resources? How could we slow down climate warming and its disastrous consequences? And what could we do to change the capitalist consumer society so that it would not flood us with unnecessary products?

For me, the EU is the world's most fascinating experiment. Is it possible to build unity without losing the cultural diversity that makes Europe so unique? Can such a foolhardy project succeed after centuries of conflict and war? I experienced the European experiment every day at EMBL. It was intriguing to observe how scientists from diverse countries behaved together under the same roof. The atmosphere was much more innovative than at home in Finland. Every nationality had its own peculiarities, which stimulated as much as they irritated. We simply lived in Europe. In my talk at the Exeter conference, I stressed how important it is that the EU promotes the quality of research and education, and I posed the question of whether it could do more by organizing competitions for new ideas to innovate research and education at the university level, which might be financed by EU funds. Despite many criticisms, the EU is delivering European solutions to important issues that we would never be able to achieve without its support. For me, this is a reason for optimism.

***

On a humbler scale, another reason for optimism was the progress our research was making. There was no *Eureka!* moment, but a conceptual breakthrough was slowly surfacing. In my notebook from June 1994, I wondered how I could have been so incredibly stupid before. We had limited our sorting concept to the Golgi complex of the MDCK cells and extrapolated our data only to other types of epithelial cells. Now, it suddenly dawned on me that the concept applied to all cell types of the body. Our concept developed into a general principle for the dynamic organization of cell membranes.

It was time to find a good name for these membrane structures that functioned in apical sorting and operated in other cell membranes as well. I fell on the idea of calling them 'rafts'. Our hypothesis was that lipid rafts float in the thin, two-dimensional fluid layers of cell membranes and provide a dynamic way to sub-compartmentalise functions in the cell membrane. The rafts are so small that they cannot be observed, even with the most modern microscopes, however, they can be activated to form larger platforms, for instance, in the formation of the membrane domains in the Golgi complex that form the apical transport vehicles. Then, they become visible. This concept of dynamic, miniscule rafts that cluster specifically to form stabilised membrane domains that function in a plethora of different activities was difficult for people to accept. My postdoc Elina Ikonen and I decided to write a review for Nature about it. Membranes are central to biology despite their neglect by researchers; now, we were introducing a completely new concept for how they are organised to perform their functions.

We were very excited to see how Nature would react. The first step went well, and our manuscript was sent out for peer review. But when the reviews arrived, we were astounded. The first referee was very positive, but the second one was brutal in their criticism. The concept was flawed, this reviewer said, but did not want to explain why. They predicted that if the review were published in Nature, it would encourage thousands of researchers to do unnecessary work. Yet, we were persuaded that our concept was novel and innovative. Provocative hypotheses such as this stimulate research because the field is incited to test them.

This time, Elina and I were lucky. The Editor, Annette Thomas, decided to publish despite the criticism, which she thought was unreasonable. Actually, I was proud that the referee had found our research so exciting that they feared other researchers would follow us blindly, like the children who followed the Pied Piper of Hamelin.

While our review was receiving such harsh criticism, I also got into a troublesome spot at a meeting in Taos, USA. David Sabatini, who worked in the same field of research as us, presented results on MDCK cells that were completely different from those I was going to present. Moreover, I was scheduled to speak immediately after him. What should I do? I remembered a similar situation at a meeting on the Italian island of Capri. Two renowned neurobiologists gave talks about the same neuroreceptor, the acetylcholine receptor, one following immediately after the other, but they totally contradicted each other. Yet, both speakers talked about their data as if the other researcher's work did not exist. I found this unscientific behaviour extremely embarrassing. They did not make the slightest attempt to explain the differences between their results. Therefore, I decided to point out during my talk that the conditions in Sabatini's experiments were different from ours and that we would have to sort out together how the divergent results had arisen. If scientists are unable to remain rational in sensitive situations, who can? I was convinced that our results were correct, but I did not want to insist.

***

My confrontations with Fotis Kafatos had made me less eager to fight for the future of EMBL, and I started thinking of other projects. Our research was in a slow phase. It was difficult to find convincing experiments to validate our raft hypothesis. Maybe I could generate inspiration by focusing my energies on some distractions. This was my peculiar way of reacting to the unbearable slowness that often dominates research. I was a member of the board of the European Cell Biology Organisation (ECBO), which arranged congresses in Europe. I offered to be the organiser of the next ECBO congress. My proposal was accepted, and thus, our team was to organise a congress in Heidelberg in 1995.

The problem with European research organisations of the ECBO type is that they are alliances of national professional societies in various fields. ECBO was an association of national cell biology societies, which traditionally took turns arranging congresses in their respective countries. The standard of the meetings was highly variable since many national societies lacked the resources to assemble an attractive programme. Also, the board members of the national societies took advantage of the chance to invite friends and mentors to give lectures when the meeting was held in their own country. I remember a meeting of the Federation of European Biochemical Societies (FEBS) in Helsinki, where a speaker was so old and decrepit that he almost had to be carried onto the stage.

The situation was entirely different in the USA, where the American Society of Cell Biology (ASCB) attracted 6,000–8,000 participants to its annual congress. It was an excellent conference and a source of inspiration for young researchers. As a speaker myself, I had experienced the great atmosphere there. Even the poster sessions were stimulating. The ASCB had a democratic organisation in which its members elected the governing board. When I became member of the ASCB board in 1997, I could see from the inside how the organization worked.

I remembered what an experience it had been for me to participate in my first European congress in 1968. It was organised in Prague by FEBS. The congress of the then-newly founded organization functioned excellently. For a young group leader from Helsinki, it was immensely inspiring to listen to well-known scientists. It was also stimulating for this young scientist to realise that not all those reputable speakers were quite as fantastic as I had thought they would be. I could even picture myself one day presenting my data to a large audience.

The 1995 ECBO meeting in Heidelberg worked out very well. However, I had no success with my hidden agenda: to persuade the ECBO board to reform the organization along the lines of the ASCB and raise the standards of ECBO's activities. I proposed that ECBO should no longer be an alliance of cell biology societies but should copy the ASCB and elect its board democratically. My ideas met with strong opposition, however, and I gave up.

Maybe the only viable solution would be to establish a completely new organisation. I even had a name for it, the European Life Scientist Organisation (ELSO). We thought it would be best to include molecular biologists from the very beginning, not only cell biologists. We all belonged together, after all. Why not accept the fact that we can learn from one another and, above all, inspire young researchers to face diversity and think more holistically?

To my delight, I was invited to deliver a Harvey lecture in New York. This was something I would never have imagined when, as a postdoc in New York, I listened to my first Harvey lecture. I went to New Haven to see my sister Majlen and her family; her husband, my old colleague Ari Helenius, was working at Yale. I combined the family visit with a lecture at Yale about my research. I was tired and uninspired. The lecture was a disaster. It was a blow to me. In three days, I was going to give my Harvey lecture at Rockefeller University. Then, everything had to go well.

I had recently given a plenary lecture at a congress for developmental biologists in Finland. I thought the presentation went well, but Erkki Ruoslahti and Antti Vaheri, my old pals from my time at Sero-Bacteriology, were of a different opinion. Erkki was now the director of a big cancer research institute in La Jolla, and Antti was a renowned cancer researcher in Helsinki. Their comments were merciless. "Kai, you stutter; so many 'Ah's and Oh's' when you speak. It just doesn't work. You must work on getting rid of those mannerisms; you need a public speaking instructor." Erkki himself had consulted an instructor who had coached Californian senators and congress members. It worked for him. Erkki's talks were much better than before. Nevertheless, I did not take their critique seriously since both were serious jokers, although I knew that I had to watch out for these defects when speaking.

The Harvey lecture at Rockefeller traditionally started with cocktails and dinner. Ari, Ira Mellman, Günter Blobel, and Jim Rothman were present, among others. All the men were dressed in tuxedos, and the ladies were dressed up as well, so it was a festive mood. All the participants around the dinner table gave speeches, mixing praise and insults, and all toasted me. The purpose was obviously to get the speaker tipsy. The atmosphere was exceptional and extremely hilarious when we marched into the lecture hall. The introduction started before I had had time to place all my slides in the carousel in the projection room. Strangely enough, the lecture went well. I am pretty sure that the 'Ah's and the Oh's' were conspicuously absent. After my talk, Jim Rothman gave the concluding speech. Remarkably, he was very flattering, although Ira and I had confronted him in our review 'The Golgi complex: *in vitro veritas*?'. Such is the world of research. We compete, but we remain friends. At least, that is how I felt then.

Chapter 5: Dresden

Would we be able to obtain convincing data to support our raft hypothesis about the dynamic organization of the cell membrane? This went on and on. Suddenly, a new project came up that moved my research priorities backstage for a while. Wieland Huttner, a former group leader in the Cell Biology programme at EMBL and now professor at the University of Heidelberg, had been contacted by the Max Planck Society (in German, the *Max-Planck-Gesellschaft*, MPG) who asked if he was interested in becoming a department director in a new Max Planck Institute in East Germany. The government had asked the Society to extend its activities to the new states of the unified Germany. The committee responsible for initiating a new Max Planck Institute for Cell Biology and Genetics had so far been unsuccessful in persuading a team of respected scientists to move East and was ready to give up. All the researchers they had contacted preferred to stay in West Germany. East Germany appeared too risky for them. Now, Wieland asked me if I was interested. He wanted to suggest my name to the search committee. "If you go, Kai, I'll go too", he promised. We would be their last resort. If I did not show any interest, then the whole project would be abandoned.

The initiative was truly unexpected. What was being planned was no small project: an institute with five departments. That would imply a team of five directors and a new building for some 400 employees.

The MPG had wanted the institute to be built in Jena or Halle. In that case, my answer would have been a flat "No!" I had seen these cities when we toured East Germany after the wall came down, and they were out of the question. Wieland had suggested Dresden, the only large city I had not seen during my East German odyssey. Dresden is the capital of the state of Saxony, and I knew it had been beautiful before it was destroyed by bombing during the Second World

War. When I was invited by the MPG committee that was responsible for founding the new institute to be briefed about the plans, we also discussed the site. I inquired whether the new institute could be located in Dresden. The answer was negative because two other new Max Planck institutes for chemistry and physics were already being built there. I did not give up, however, but asked if I might be allowed to work politically for a Dresden solution. This was possible, but I was told that I must not contact the state government of Saxony, which would certainly do everything in its power to scoop up a third Max Planck institute.

Naturally, no one thought that a Finn could have political influence in Germany. The Max Planck institutes are well funded, so there is often a political struggle for the location of a new institute. The responsibility felt heavy. If I declined, this central field of research would not get the reinforcement it needed in East Germany, and I had seen for myself how necessary such an investment was. Enthusiasm at home initially was not great either. Katja, Mikael, and Matias were already studying at university, which meant that only Carola and I would move if we settled for Dresden.

Before deciding, we wanted to see the city. The whole family went to visit. We liked it. If the offer to move had come up earlier, we would have been dismayed at the decay, but what we saw now was a city being rebuilt. We could imagine how beautiful it would be once all the building work underway was completed. The surrounding countryside with its woods and mountains was beautiful too, and the city was known for its active cultural life. Also, thousands of villas in the city were being restored. These villas, surrounded by large or small gardens, were typical of Dresden. They had been built during the economic upswing at the turn of the nineteenth century known as the *Gründerzeit*. Despite their dilapidation, one could see their quality and beauty.

The Dresden project was looking more attractive, but I was still hesitating and not sure I wanted to commit myself to such a huge undertaking. I was approaching sixty and had already participated in building the Haartman Institute in Helsinki and EMBL in Heidelberg. I had been demonstrably useful to both projects. Did I really have to take on another project to prove myself?

Or maybe I should accept it precisely because of my previous experience? My interest in projects that involved teamwork was aroused by my key experience during my school days in Ågeli. Even then, I realised how much better everything goes when you work together towards a common goal. Above all, it could be great fun and gratifying to collaborate with others. Would it not be rewarding to try my hand at one final project? Would it be possible to build a top-level institute in a city where there was no molecular biology? The Dresden *Technische Universität* (TU Dresden) was a classical technical university. Its medical faculty had been founded only a few years earlier. In practice, we would have to start from zero. I remember a Max Planck director friend asking me how I could even consider moving to Dresden where we would be alone in a city with no molecular life sciences. Indeed, there was practically none and just because of that, the project was a mightily exciting challenge. We had the advantage of starting afresh.

***

Western society has long lacked an understanding of cooperation as a fundamental organisational principle. Instead, the individual gets all the attention. In this respect, the MPG is no exception. It consciously follows the principle of Adolf von Harnack, the first president of the Kaiser Wilhelm Society, the predecessor of the MPG before the Second World War. Harnack's Principle entails selecting strong individuals with excellent research records as Directors of the departments in their institutes and giving them ample funding until retirement. It was the antithesis of what I believed in.

Over the years, I have tried in vain to understand why cooperation and altruism play such an insignificant role in politics and economics. Most of us have positive experiences of togetherness and collaboration. People have always worked together. Nonetheless, our societies are dominated by strong individuals, and mostly men. The notion of social Darwinism introduced at the end of the nineteenth century initiated an era with devastating consequences for our planet. Today, our societies are mired in consumption and gluttony, and individualism reigns supreme. In this selfish world, altruism has been pushed aside as a principle with little relevance to modern life.

One of the most influential proponents of brutal egocentrism was writer and philosopher Ayn Rand. She wrote two novels that became incredibly popular, mainly in the USA. Both had selfishness as their theme and lodestar. Rand's books have sold twenty-nine million copies. Her 1943 novel 'The Fountainhead' ranked second, just after the Bible, as the most influential book in the USA in a 1991 poll. Another of her books had the title 'The Virtue of Selfishness'. Ayn Rand's theses greatly influenced politicians, especially in the USA and the United Kingdom, and provided moral support for neoliberal economics. Alan Greenspan, Chair of the US Federal Reserve for many years, belonged to Ayn Rand's inner circle. Egocentric rationality was her mantra. "A man's ego is the fountainhead of progress," is one of her famous sayings. Ayn Rand, without a doubt, would have accepted Margaret Thatcher's claim that there is no society, only individuals and families. Economic theory has been entirely infused with the idea of self-interest as the guiding principle of humankind. *Homo economicus* is a rational, strictly calculating, and selfish individual. This half-monster caricature of human beings has been used by economists as a model in their algorithms. No wonder we have become such individualists.

There are other points of view, however. Piotr Kropotkin, for instance, promoted the principle of altruism, so hotly detested by Ayn Rand. I first read his autobiography 'Memoirs of a Revolutionist' (translated into Swedish as 'Memoirs of an Anarchist') and later his book 'Mutual Aid: A Factor of Evolution'. Kropotkin was born in 1842 into a princely family in Moscow, Russia. He was educated at the Imperial Page Corps in St. Petersburg, which gave him the freedom to choose his professional career. To everyone's surprise, he chose to join the Siberian Cossacks. He was interested in Darwinism and wanted to investigate whether the struggle for survival existed in the wilds of Siberia. Kropotkin reached a completely different conclusion to Darwin from his studies of animal life. He observed that the animals in the hard climate of Siberia did not fight one another, but instead cooperated. Russian biologists had reached the same conclusion. They held Darwin's theory of evolution in high regard but emphasised cooperation more than competition and drew conclusions that were quite different from those of the Social Darwinists. One reason for the differences between the findings of Charles Darwin and Alfred Russell Wallace may be that the Englishmen studied the animal world in tropical regions with high densities of individuals. In the cold climate of Siberia, populations were much smaller. 'Mutual Aid' was published in 1903 but it had no impact whatsoever. Darwin himself took altruism seriously, but his successors cultivated the notion that only the strongest (or fittest) survive. This became the leading principle in the British and American societies of the late 1800s, and it has marked the culture of our Western societies ever since.

On my many visits to scientific institutions across the world, I had seen how research was hampered by individual conflicts and poor cooperation. Researchers often fought about whose name would be first and whose would be last on their articles. All the names between these were mostly regarded as lesser, 'sandwich authors'. Often scientists accused one another of stealing ideas. The ideas were rarely so important that they deserved to be fought over but, still, they poisoned the atmosphere and blocked cooperation within the institutes.

After long deliberations, I started to wonder whether it might be worthwhile to accept the challenge and try to build a new Max Planck institute on the principle of cooperation rather than the Harnack Principle. If we managed to build a top institute programmatically based on mutual support, we might also become a model for others. I began to feel enthusiastic about the project. We had to set the bar high from the very start; otherwise, the project would not merit the effort. Such an ambitious undertaking would require an immense commitment from me. As usual, I was met with scepticism when I discussed plans with colleagues, but my experiences from Helsinki and Heidelberg made me hopeful. The all-important prerequisite, however, was to assemble a dream team that could be persuaded to move to Dresden.

***

Now, we had to get started. I contacted Konrad Müller, who had recently resigned from his position at EMBL; he immediately promised his support. Our first task was to ensure that the new institute would be located in Dresden.

Konrad activated his network. He knew Kurt Biedenkopf, the Minister President of Saxony. Biedenkopf had been a rising star in the Christian Democratic Party but had been pushed aside by Helmut Kohl, who did not tolerate competitors. I had been forbidden to contact Biedenkopf, but that restriction did not extend to Konrad. Konrad also knew Bernd Schmiedbauer, who was number three at the Federal Chancellery (the office of the German federal government) under Kohl. Schmiedbauer was happy to accept the task of creating political support for Dresden. I heard that the Max Planck leadership was not particularly liked in the Federal Chancellery. MPG presidents were often arrogant and went their own ways. They insisted upon the freedom to carry out their research and did not bend to the wishes of politicians. The MPG stood for basic research, full stop. Germany was lagging behind the world in biotechnology, however, and the government thought that this field had to be supported in East Germany as well as in the West.

My next task was to select a team that would build the scientific structure of the new institute. Wieland Huttner was already on board. If we were to succeed in creating an organisation built on cooperation, we ought to know our team of leaders inside and out. The task was so difficult that we could afford no unpleasant surprises. Some people are simply incapable of cooperating. The team should have a strong international foundation, just like at EMBL, where we learned how mixing scientists of different nationalities enriched the research environment. I decided to ask Tony Hyman and Marino Zerial, both group leaders in Cell Biology at EMBL. Tony was an Englishman and Marino an Italian. Diversity promotes creativity.

Marino was cautiously positive, but Tony initially showed no interest. For me, he was essential for our quartet because he was the youngest. I wanted young people involved in building the institute. Tony was a rising star and was already getting offers from other institutes. He had done his thesis work at the LMB in Cambridge and his postdoc at the University of California at San Francisco, one of the most influential institutions in American molecular biology. Tony already had a network of important contacts. Also, Suzanne Eaton, his spouse, was a group leader at EMBL and an outstanding researcher. She had done her postdoc in my lab. Here was our dream team. I just had to get them all onboard the project.

We needed an attractive and convincing research programme. We wanted to combine the successful cell biology that we had built up at EMBL with developmental biology and physics. We decided to focus on how cells form tissues. The fusion of cell and developmental biology with physics was a venture that had not been tried yet. Our slogan would be 'How cells form tissues'.

This ambitious research programme would need an extensive infrastructure, including modern microscopy, mass spectrometry, and other methods and equipment. We could no longer concentrate on cells in culture but had to start using model experimental animals: fruit flies, worms, fishes, and mice.

Our most revolutionary idea was to do away with classic Max Planck Institute departments because they created barriers to communication. The five departmental budgets planned by the MPG were fused into a single big institutional budget. And the budgets of the directors were smaller than planned, so their research groups would be small too. The resources thus liberated went into building a common infrastructure, and funding twenty research groups led by young investigators with fixed-term contracts. We simply copied the model that was tried and tested at EMBL.

The EMBL-like concept of our institute was totally new for the MPG and for most of Europe. Marino, Wieland, and I went to Frankfurt to present it to the MPG committee in charge of founding the Max Planck Institute for Cell Biology and Genetics. This committee, which consisted of directors from other Max Planck institutes in biology, had worked for four years without results. To our great surprise, leading members of the committee embraced our proposals enthusiastically. Less surprisingly, we also met with some resistance: the structure we proposed for what we were calling the MPI–CBG was considered too radical. But all in all, we returned to Heidelberg optimistic and happy.

We were met by Konrad, who had a joyful message: Schmiedbauer at the Federal Chancellery had told him that Helmut Kohl advocated for Dresden as the location of the MPI–CBG. When the Chancellery recommended Dresden as the site for the new institute, there was no further opposition to our plan.

Next, the biological–medical section of the MPG, comprising all institute directors in that sector, would process our proposal and evaluate our team. Each of us would be assessed by at least ten international experts. We waited in suspense for the results. I had made it clear to the MPG that both our plan for the organisation of the institute and the acceptance of all four members of our team – Marino, Tony, Wieland, and myself – were essential. We would all withdraw from the project if these conditions were not met.

The section met and gave us a positive response. They had some concerns about Tony because of his youth – he was only thirty-five. But, ultimately, they approved him too. All that was left to do was to persuade Tony and Suzanne to join us. For me, this was phenomenal. The directors of the MPI–CBG would span the age range from sixty to thirty-five. We would engage twenty young group leaders. Youth would dominate our staff, and this was, in my experience, the basis for success.

We planned to stay at EMBL until we could move into our new building in Dresden. An interim arrangement in Dresden might have killed our project. Too many problems might have arisen that we could not control. When I notified Fotis Kafatos about our plans, he became extremely upset. He threatened to fire me immediately and to split the Cell Biology programme into project groups which he would lead himself. What an absurd idea! His violent reaction gave Suzanne and Tony such a scare that they quickly decided to join us in Dresden.

Gradually, Fotis calmed down. Cell Biology at EMBL had recently been evaluated by a group of external experts who had given us a fantastic report. The group warned Fotis that he would have problems with the member states if he decided to discontinue the programme. The MPG agreed to pay all our costs at EMBL until we moved to Dresden. This offer was timely for Fotis because the budget of EMBL was tight at that time. The MPG also accepted to pay the costs of everyone we hired during the interim period, even researchers working at other locations. The Society did all in its power to pave the way for a successful start of our project. This was indeed very welcome support. The flexibility of the MPG was astonishingly positive.

The first visit to Dresden by Konrad, Marino, Tony, Wieland, and I went well. The city was still full of decayed buildings, but the restoration was making good progress, and new housing was being built. We had lunch in Pillnitz Castle on the river Elbe. We saw paintings by Raphael, Vermeer, Rembrandt, Canaletto, and other old masters in Dresden's baroque Zwinger. Part of Dresden's former glory was still intact. Even Tony was positive. In the evening, we had dinner with the Max Planck committee, high officials from the ministries, and city representatives. Everything was jovial and relaxed. The following day, we visited TU Dresden. Here, too, people were incredibly positive and amiable. We inspected possible sites for our new institute and found an ideal location next to the Medical Faculty and the University Clinic, 200 meters from the Elbe. In the evening, we went to the Semper Opera and saw Mozart's *Le Nozze di Figaro*. When we returned to Heidelberg, we were firmly determined to realise the project.

Shortly after our return, I received a letter from the Mayor of Leipzig inviting Carola and me for a long weekend in the city. He wanted us to acquaint ourselves with Leipzig in the hope that we might possibly change our plans. I had already seen Leipzig and was not terribly interested, but, to be polite, we accepted the invitation. The weekend was both pleasant and interesting. The city did everything to impress us. But there was no rapport with the university, which was much less welcoming and interesting than the university in Dresden. During the dinner, Carola and I were completely excluded. We sat alone in a corner of the table while the rector entertained his professors. I have never experienced anything similar. The impression was that they certainly did not want us around. The message we heard was "Do not disturb our circle." After the visit, we thanked the mayor and his wife for their great hospitality, and, later, we informed him regrettably that we had decided on Dresden.

The negotiations with MPG President Hubert Markl and the MPG went on. Our quartet had divided the tasks so that each of us had clearly defined obligations. Wieland was mainly responsible for our finances. After a while, I received signals from the members of the MPG committee supervising us that we had reached the upper limit of what we could ask for in funding. The MPG could not raise the budget to the level envisaged by Wieland. Nevertheless, Marino, Tony, and I thought it was time to sign our contract. We did not want to put up a fight that could cause us trouble in the future. But Wieland remained stubborn. The MPG held its annual meeting in Berlin, and I called the committee to inform them that we were ready to sign. But I had to tell them that Wieland was not yet ready to join.

One week later, the full budget was approved. Wieland had won! Only later did I realise how brilliant his strategy had been. The MPG could not risk that the only German on the team would leave. The committee knew how difficult it was to recruit top German scientists to East Germany and had no option but to accept Wieland's demands.

The plans for our new institute had now advanced so far that we were ready to negotiate our conditions of employment with Markl. Normally, the MPG president negotiates with each candidate separately, but I insisted our quartet should negotiate together to emphasise that we intended to work as a team and to justify our new plans for the institute. I hoped that my recent election as a foreign member of the USA's National Academy of Sciences would have a positive influence on the outcome.

Marino, Tony, Wieland, and I travelled to Munich together to a meeting that we shall never forget. Our unconventional demands had probably already infuriated Markl. We did not conform to the scheme that the organization had built up; we wanted to break with the MPG structure that was too heavy with tradition. But despite the unpleasant atmosphere, Markl had no means of stopping us. We had the blessing of the highest echelons of government. By and large, Markl had to accept our demands.

Another problem at this stage was the role of Konrad. So far, I had paid for his contributions from my own pocket. Naturally, he could not work free of charge. The MPG administration now said that they wanted me to stop using Konrad's services. I believe the president was irritated by Konrad's contacts in high places.

I was forced to negotiate with the administrators in Munich, who still refused to pay the full sum that Konrad asked for. I insisted that we would continue to use Konrad as advisor in the future, too. Since we had not yet signed our contracts, I could practice a certain degree of extortion.

For me, it is still a conundrum that the West Germans did not understand the fantastic opportunities being offered in the East. Innumerable professorships were vacant at the Eastern universities. Few established West Germans applied. Not even the young, rising stars understood the exciting prospects in the East. This was now the place where you could realise new ideas. Marino, Tony, Wieland, and I signed our contracts on 6 December 1997.

***

Now we had to find suitable architects for the new building. My brother Tom was Dean of the Department of Architecture at Helsinki

University of Technology, so I asked him to suggest a good candidate. I wanted a Finnish architect so it would be easier for us to negotiate about the design of the building.

I have always been amazed by the architecture of research institutes. To put it bluntly, they are almost invariably built to hinder rather than facilitate communication. How often have I been lost in buildings constructed like a maze? In our new institute, even the architecture should facilitate cooperation. Normally, the government required a competition to select a suitable architect for a building project of this size, but since the MPG was erecting more than fifteen new institutes in eastern Germany in only a few years, this obligation was waived.

Tom suggested Mikko Heikkinen and Markku Komonen, two leading Finnish architects. I contacted them, and they were interested. Marino, Tony, and I decided to make our own sketches to show the architects how we imagined the building. We assumed that it would be built on the plot on the river Elbe that we had seen, and we started to plan accordingly. It was exciting to brainstorm together. It was also important to air our ideas about how work at the institute should be organised. With the sketches, we wanted to illustrate our thoughts about communication within the building and what the labs should look like. Should we have labs designed for single groups or larger spaces shared by several groups? Also, it was important that the architecture should have that *Wow!* factor. We wanted to impress our visitors, especially those we wanted to recruit, because we understood that it might not be easy to entice researchers to move so far east. Most importantly, the architecture should promote collaboration in the building.

When our plans were ready, I went to Helsinki to meet Heikkinen and Komonen. They were surprised that we had gone to such lengths, and that was exactly what we had hoped. We wanted to encourage them to take on the project. They had not planned a research institute before but thought it would be exciting to work with us and accepted the task.

It was time to convince the building department at MPG headquarters that their architect, Günter Henn of *Henn Architecten*, should collaborate with Heikkinen and Komonen. Henn had an enormous office in Munich. He was also Volkswagen's trusted architect and had planned the city of Wolfsburg in northern Germany, which was built in the first half of the twentieth century as a home for workers in the car industry. Henn was not very interested in working with the Finns, but we insisted. The MPG decided that each architectural firm should draft its own plan for the MPI–CBG building, and then there would be a decision about which design to choose. In any case, Henn would have to lead the construction work since, by law, that task could not be assigned to a foreign company. Now, the question was whether the plan of Heikkinen and Komonen or that of Henn would be victorious.

When the plans were ready, we were all called to Munich to decide how the project should proceed. Both plans for the MPI–CBG building were presented. We had continued the collaboration with Heikkinen and Komonen and were excited about their draft plan. We needed to present our views in such a manner that they would get the commission. Fortunately, we had a new argument: a copy of an article in The New York Times. The duo had designed the Finnish Embassy in Washington DC, which had become something of an icon for architects in the USA. The Times article was overwhelmingly positive. The full-page spread under the headline 'The Finns go for Baroque' was a perfect link with the baroque city of Dresden.

Without insulting Henn, we did our best to support the Finnish proposal. Naturally, we stressed that it would be an exciting collaboration. Fortunately, the director of the MPG's building department understood the architectural merits of Heikkinen and Komonen's plan. Heikkinen and Komonen won the commission to design our institute together with Henn. Afterwards, Henn came and congratulated us. He was Professor of Architecture in Dresden, and I had quite a lot to do with him later. He was a good loser. Actually, he was no loser at all. Both offices learned a lot from one another. Collaboration pays off.

There were many reasons to be delighted with our project. One was the location of Dresden. The city lies right in the middle of Europe, with Poland, the Czech Republic, Austria, and Hungary as neighbours. During the years of the Iron Curtain, there was only East and West Europe, but before the Second World War, the region around Dresden was Central Europe.

In the nineteenth and early twentieth centuries, Central Europe was a flourishing centre of culture and science. The Habsburg Empire encouraged creative personalities, irrespective of their ethnic background or religion. The open policies of the Empire attracted talents who blossomed in the stimulating cultural environment. All this ended with Hitler and his fascist regime.

We will probably never recover this pre-war Central Europe. For one, the strong Jewish component is gone forever. Nevertheless, Central Europe is back on the map of Europe; it is much more provincial than before but has great potential. It was this potential we wanted to activate by creating a new research environment in Dresden.

We imagined that we would be able to create a collaborative network of molecular biology research institutions comprising the Central European cities of Warsaw, Prague, and Budapest in addition to Dresden. They needed us since their molecular biology was as weak as that in East Germany. Backed by the MPG, Wieland and I set up a partnership with an international research institute in Warsaw that was supported by UNESCO. A German group leader, Matthias Bochtler, went to work in Warsaw, and a Polish group leader, Eva Paluch, came to the MPI–CBG.

We managed to get this partnership running and thought that Warsaw would be the first node in our network. Then we got stuck. Unfortunately, Wieland and I could not convince the Research Ministry in Berlin that an investment would be profitable. That was a real pity. The national borders remained barriers, even after the countries had joined the EU.

***

Our daughter, Katja, had begun to study sociology in Berlin. When I first visited her in Berlin-Mitte, which was in East Berlin before reunification, the decay was conspicuous. The façades of the old buildings were still marked by bomb damage from the end of the war. I accompanied Katja to a discussion in the district of Prenzlauer Berg about the situation in East Germany seven years after reunification. The people on the stage were top East German politicians, including Gregor Gysi and Wolfgang Thierse. The discussion revealed how difficult it had been for most East Germans after the fall of the wall. It was also interesting to observe the reactions of the listeners. It will take a long time before the wounds heal from the fusion of two completely different social systems. Helmut Kohl was right to push for unification and the dissolution of the GDR without delay, but what happened next was poor planning and politics.

West Germany transferred its social system to the East mercilessly and, in doing so, practically demolished East German society. The country was indeed bankrupt, but there were features from the GDR that might well have been introduced into the West. For instance, the GDR had a much better system of childcare. Also, their health centres and out-patient clinics would have been worth further development. On top of that, West Germans were openly arrogant towards East Germans. The humiliation the East Germans experienced was a recurrent theme in the discussion in Prenzlauer Berg. The Kohl government had to dismantle most of the GDR's industry because there was no market for their products after 1990. Not even the East Germans wanted them anymore. This revolution was carried out without transparency, however, and without explaining why the GDR was on the brink of bankruptcy. Their government had never informed them about the true state of their economy. It was impossible to build a new economy before the old one was scrapped. Yet it was both irresponsible and unwise to unleash this enormous change without explaining why no other solution was possible. Why was there not more empathy for the East German people in this unique situation?

Katja and I could discern that the political views of young people in the audience in Prenzlauer Berg were moving to the extreme right. Thus far, the problem stayed under the surface but, in the long run, there would be trouble.

***

During all this manoeuvring, research in my EMBL lab continued. We were in a phase that required new approaches, so my presence was important if we were to be successful. But I simply had too little time to devote to our research projects. During this hectic period, my family helped to maintain the productivity of the lab. I was collaborating with our son Mikael, who studied medicine in Heidelberg and was now working on his doctoral thesis. Mikael worked on Alzheimer's disease and had shown that our rafts played an important role in formation of the protein beta-amyloid, which is a component in the pathogenesis of the disease. Interestingly, Mikael could inhibit the production of this protein by lowering the cholesterol content of nerve cells.

Mikael had followed my example in combining medical studies with research. He became a neurologist and researcher. Our other son, Matias, also began to do research during his medical studies and, likewise, became a researcher, but he didn't specialise as a physician.

Katja defended a doctoral thesis in sociology and later moved to New York, where she was engaged in work for German research organizations and universities.

At home in Heidelberg, Carola and I lived alone in our townhouse, as all the children had departed. Carola's dental office kept her fully occupied. She has always been a fantastic partner. As usual, she was foresighted and pragmatic. Before the move, her only demand was that I should stop smoking my cigarillos. I smoked one, sometimes two, each day when I came home, while I looked back at my day and contemplated what I had on my plate and what I ought to do next. I also imagined that I could get strokes of genius while I smoked. But Carola hated the stench of tobacco smoke. I found I had better obey, stop smoking, and find other ways of planning my day. Anyway, my best ideas always came to me during discussions with others.

Now Carola and I would move again, for the seventh time in our married life. I have often pondered how peoples' identities, especially children, are affected by moving from their home country. In a television interview in Finland, I discussed precisely that question. Are you Germans? No, definitely not, we are Finns. We still have Finnish passports. Finland remains our homeland even if we live in another country. Besides that, we call ourselves Europeans. This was a new element in our identity and one we felt strongly about. After the interview, I received e-mails from several Finns in the same situation. Like us, they had pondered their identity and were happy with the concept of being European. I think it is fantastic to be able to combine your origin – Finland for us – with your present location – Germany for us – into an identity of being European. This is why the European experiment is so terribly exciting for me. Will we, Europeans, be able to pull it off successfully? I hope so!

***

We started to plan the relocation of the lab to Dresden, but it was not yet imminent. I had followed Ari's problems when he moved from Yale to the ETH in Zürich to become Department Chair of Biochemistry there: he hadn't been able to persuade anyone from his lab in the USA to move with him. He did manage to build a new research group, but it cost him lots of work and time. I could not afford anything like that. I needed to bring accomplished and reliable co-workers with me to Dresden. My most successful postdocs did not want to join me. They applied for positions to continue as group leaders elsewhere. Therefore, I started working to convince a few selected members of my group to join me in building our new lab in Dresden.

At the same time as we were busy planning the Dresden project, we had decided to found our new society, ELSO, and were preparing our first ELSO meeting, which was to be held in Geneva in 2000. I had been elected President of ELSO by a founding committee with Paul Nurse as chairman. His authority as an influential British Nobel laureate gave credibility to the project. The composition of the committee reflected molecular biology in Europe. ELSO would be an organisation not just for cell biology but would represent molecular biology in its entirety. This was our hope. Now, the big issue was whether we would succeed in attracting enough participants to make ends meet. We assumed there was a need for the type of congress we were preparing. There was not a single major meeting in Europe that came close to the quality of the ASCB meetings. We hoped that many of those who usually went to the USA to attend the big American congresses would now choose to participate in an ELSO meeting.

Time to get started. We had an economic surplus from the successful ECBO conference, but we needed more resources. Our ELSO team was small, but efficient. Ingeborg Fatscher, who had managed the ECBO meeting, handled the practical arrangements.

Konrad was appointed General Secretary of ELSO, and I hoped that he would manage to find more funding for our congress. Although we were aiming to attract two thousand participants, we did not want to use a professional congress organiser. Neither did we have the resources to employ a larger staff. We decided to do everything ourselves. In that way, we hoped to recreate the cordial atmosphere of the Heidelberg congress. With a professional organiser, everything is so anonymous. ELSO had also become an important element of the Dresden project: the next congress after Geneva would be held in Dresden in 2002. That would give us the opportunity to entice researchers to visit Dresden and see for themselves how attractive the city was. The former Eastern bloc was still viewed with great suspicion in the West. We had to overcome that.

We managed to persuade the Swiss Biochemistry Society to hold its general meeting together with ELSO in Geneva. This increased our confidence. Once again, our extensive network proved helpful. Two group leaders from Cell Biology at EMBL, Jean Grünberg and Thomas Kreis, were now professors in Geneva. Now we would have to compose a programme attractive enough to bring enough participants to the congress. The first ELSO meeting would have to be a success.

We also hoped that ELSO would become a mouthpiece for European researchers in the EU. The EU had become an important source of research funding, and their programmes allocated substantial budgets to support research. Because of my position at EMBL, I had been active on various EU committees in Brussels and did not want to lose my foothold in the EU bureaucracy; it was necessary both for the sake of Dresden and our high-flying plans for ELSO. I was also chair of the commission that evaluated the fifth EU Framework Programme for the Life Sciences (1994–1999). It was a gigantic task, but we focused on how the researchers evaluated the programme. We sent out a questionnaire that gave researchers the opportunity to present their own suggestions for improvements and ideas for the next framework programme. The EU's selection of research fields to be funded generated the most criticism. The fields were sometimes so narrowly defined that only one or two research groups were eligible. Such formulations could only arise through insider contacts. The EU did not mobilise a sufficiently broad expertise to participate in the planning.

Most importantly, the commission unanimously proposed that the EU should launch an initiative to make it easier for young researchers to start an independent career. There, the USA was miles ahead. We proposed a new programme, Euroexcellence, focused on young investigators. The programme would have an annual call for young researchers, where the best applications would be awarded five-year research grants. The decisions would be made by committees with relevant expertise from all member states. For me, it was clear that such a programme would bring a revolution in European research. If the lethargic professorial hierarchy could be challenged by young investigators with their own grants, barriers would fall at a single stroke. We wanted to eliminate the situation where university professors filled positions at their institutes with their henchmen. This caused much of the best European talent to stay in the USA after their postdocs. This drain had to be stopped. My optimism knew no boundaries, and that gave me unexpected strength.

***

We still had a fifth Director's position vacant at the MPI–CBG and initiated a serious search for candidates. The position was initially reserved for a developmental biologist, but Tony came up with a brilliant idea. Joe Howard, a fantastic biophysicist, had spent a sabbatical in Heidelberg, and we all knew him well. He was a professor in Seattle, but originally came from Australia. We ignored the recommendations of the MPG committee and put our energy into persuading Joe to join us. We also wanted to attract his spouse, Karla Neugebauer, who was an up-and-coming cell biologist and a perfect fit for the MPI–CBG. Joe and Karla came to visit Dresden and immediately clicked with the team. The MPG accepted Joe without problems, and our quartet of Directors became a quintet. Joe was an enormous asset to our project. Finally, we were close to realising our dream from EMBL: establishing a research programme in the physics of life.

The next step was persuading Peter Fulde, Director of the Max Planck Institute for Physics in Dresden, to include biology in its research programme. Fulde was a renowned physicist and the only West German Max Planck Director to have moved East. His institute's focus was on the physics of complex systems, which meshed perfectly with the physics of life. Fulde entrusted us with finding a suitable candidate to lead the new biology programme at his institute. Joe proposed Frank Jülicher. Frank came from the Curie Institute in Paris, where a strong group worked on the physics of life. Frank quickly became a mainstay of biophysics in Dresden.

Our original plan was to contribute to the development of a new research landscape in Dresden. If we allowed our institute to become an ivory tower, we would perish. Molecular biology was neglected at TU Dresden, as at many other universities of technology all over the world. Engineers knew little about biology. We wanted to change that. We proposed that a new research institute be founded in our neighbourhood. We drafted a plan for it, including six new professorships at TU Dresden in bionanotechnology.

All living cells are stuffed with machines that perform work. These machines are incredibly versatile. They transform energy, function as sensors, and produce materials of the most varied kinds. All natural products are produced by this machinery. Just think of silk, cotton, and wool. The bioengineers of tomorrow might feed their imaginations with ideas from biological research. Nature is a unique resource for new technological solutions. The potential of these solutions is practically unexploited.

The machines of the cell are minute; they work on a nanometre scale. This nanoworld should be accessible to a new generation of engineers. The machines of the cells might be our salvation and help us to make our technological base sustainable and energy efficient. A whole new world was opening up to the engineering skills of *H. sapiens*.

With this plan for TU Dresden, we approached the Minister President of Saxony, Kurt Biedenkopf, and argued ardently for a Bioengineering programme. I told him that we had already founded a biotech company, Cenix, in Heidelberg, which we would bring with us to Dresden, and that we planned several similar start-ups. The new research building we proposed for TU Dresden would, therefore, need sufficient space for commercial projects and companies in the field of biotechnology. Our slogan was 'Research and commerce under the same roof'. Günter Blobel, a founder of cell biology who was awarded the Nobel Prize in Physiology or Medicine in 1999, became an important spokesman for our Dresden project. He had donated his Nobel Prize money to rebuilding the Dresden *Frauenkirche* (the Lutheran church in the centre of the city, which was destroyed in the war) and building a new synagogue. Blobel promoted us wholeheartedly. We could not get more prominent support.

Biedenkopf and his Minister for Research, Meyer, were also on board. Biedenkopf recommended that we contact the rector of the university to ask him to include our project in his plans. We contacted Rector Mehlhorn, who viewed our plan positively. The project took off before we could quite understand that it had gone so well.

To put even more wind in our sails, I started to draft a syllabus for masters studies in Molecular Bioengineering with Jonne, Ari and Majlen's son, who now studied bioengineering at the University of

Michigan. We finished the programme with input from external experts. After incorporating feedback from our quintet of MPI–CBG directors, we submitted the draft to the Rector. He appointed a committee to develop the proposal further, and we waited for the result.

By chance, I was in contact with Francis Stewart, who was a group leader in the Gene Expression programme at EMBL and an expert in genome engineering. He would be a perfect fit for Dresden. I told him about our plans for molecular bioengineering in Dresden and the new centre we were founding. Francis was interested immediately, especially because the project included teaching engineering students. The only problem was that he had just been offered a professorship in Nijmegen in the Netherlands, where a new institute for genome research had been founded. Nijmegen was a tough competitor, but I did not give up. In Nijmegen, everything was already in place, whereas our project was still in the planning stage. Maybe it was precisely for that reason that Francis became increasingly interested. He asked what we could offer him. Nothing had been fixed, I admitted, but Biedenkopf had promised funding for the project.

Francis had to give his reply to Nijmegen, and a miracle occurred. Although he had a family with four children, he decided in favour of Dresden. Only an Australian could have made such a brave decision! He trusted us. Now, we would have to ensure that our plans for six new professorships were realised. With Francis on board, it would be much easier to recruit the others. He became our stamp of quality for the whole project.

***

New opportunities turned up everywhere. The German Ministry of Research announced a competition called Innoregio for innovative projects aimed at building new industries in all fields in the former East Germany. It was meant for big projects with price tags up to forty million D-Marks (about twenty million euros). In Heidelberg, I had participated in a similar competition called Bioregio, which focused on biotechnology. Heidelberg was one of the winners of the competition, and I had seen what positive consequences such a big project could have for a region. Thanks to Bioregio, Heidelberg grew into a commercial centre for biotechnology.

With Innoregio as a carrot, we started to analyse the commercial potential of the Dresden region in the field of biotechnology. When we contacted the Saxon State Ministry of Economy in Dresden, we were disappointed by their lack of support. They thought that Dresden's chances were small since the city simply had no modern biotechnology. The competition would be hard, but we persisted.

The MPI–CBG had engaged a resourceful Austrian, Markus Mikl, as our public relations officer. Markus and I sat down to plan a proposal for Innoregio. Officially, the institute could not be the applicant, but the University had an organisation for commercial projects, *Gesellschaft für Wissenschafts Transfer* (GWT), that was willing to take on the commission. But we had to write the proposal ourselves.

Since we could not build our project on existing biotechnology companies, we presented future plans, among them our plans for molecular bioengineering. We already had one start-up to show and planned new biotech enterprises in Dresden. Francis had founded Gene Bridges in Heidelberg, Tony had founded Cenix, and Marino and I were planning to launch another start-up, Jado, with Temo Kurzchalia, who was now a group leader at the MPI–CBG.

The proposal was completed in time, and we sent it to the GWT for submission to the National Research Ministry in Berlin. For safety's sake, I called the GWT on the due date for the application to check that everything was in order. That was more than fortunate because no one had taken care of the matter. Our application was lying on the desk in the room for incoming mail, under a pile of other letters.

Unbelievable! I had to mobilise our driver in Dresden, who fetched the application and drove to Berlin at express speed.

What would we do if the application were approved? We would face another mammoth task. We would need more personnel, both to manage such a big project and to support the founding of new biotech start-ups. Moreover, at the MPI–CBG, we had no one who could build up the necessary infrastructure that we planned for our institute. Where might we find the right people for such demanding jobs?

Once again, our networks provided a solution. Tony had studied alongside the accomplished cell biologist, Ivan Baines, in London. Baines now worked at the USA's NIH and had switched to administrative tasks. NIH distributed thirty billion dollars per year to biomedical research. Ivan had worked both at this funding agency and on an NIH commission that financed biotech start-ups.

Ivan embodied all the skills we needed. Luckily, he had grown tired of the USA and wanted to return to Europe. We quickly invited him to meet us in Heidelberg. The remaining questions were how to persuade him to come to Dresden and, above all, how we might pay him a competitive salary. With his experience, Ivan would have no difficulty finding a well-paid job in industry. But, again, a miracle happened. Ivan found our project so exciting that he decided to join us. He became Chief Operating Officer of the MPI–CBG. The quintet of directors was now a sextet!

The first thing Ivan did was to plan the infrastructure of the MPI–CBG. An entire floor would be reserved for the service units that would provide the technical support our research needed. Normally, complicated equipment was procured by the leading researchers in a research institute and handled by the scientists in their research groups. Centralisation of the technical facilities made research more effective, more professional, and less expensive. Few research institutes had anything similar. The service was to be available to the whole institute, which meant that all our research groups would have access to facilities maintained by technical experts. The service units became a resource that made our institute more attractive, and became a pillar for support of our research agenda.

***

These were hectic times for us all. Not only were there our professional concerns, but also we had to prepare for our personal moves to Dresden. Carola and I started looking for a new home. The choice was fairly large, but we wanted to live within walking distance of the MPI–CBG. In 1999, we found a villa that met all our expectations. It had been built in 1865 in Blasewitz, near the institute, in the Italian style, with Lady Justice adorning the façade. The villa was well-built but totally dilapidated. It was surrounded by an overgrown garden, which was, however, small enough for us to manage. We could not have hoped for anything better; we just had to go for it. Prices in Dresden were still affordable, although it was difficult to estimate the renovation costs.

One month after we purchased the villa, we got a shock. We received a letter from a bank informing us that our villa would be auctioned in a forced sale! We contacted our notary to learn what had happened. In Germany, a notary costs a fortune, so we wanted to know what had gone wrong. The notary calmed us by saying that everything was in order. In East Germany, sorting out ownership took a long time. Later, we heard about swindlers who sold houses they did not own or were in collusion with criminal notaries who endorsed the contracts. When the wall fell, the East Germans were exposed to crooks, who tricked them into taking insurance policies they did not need or sold consumer products at excessive prices. This was one of the reasons for East Germans' frustration. Few politicians understood how different the systems were in East and West Germany. The tsunami that rolled over East Germany left permanent marks.

We engaged a Dresden architect to renovate our new villa. Carola was responsible for overseeing the work, which she did mostly by phone from Heidelberg, but she also drove to Dresden to check the progress. There were many problems to solve and decisions to make. We hoped that all would be ready before our move in February 2001.

Konrad knew the billionaire Klaus Tschira, who lived in Heidelberg. Tschira was one of the founders of the German software company SAP, which had grown into a global megacompany for business systems and book-keeping. When Konrad told him about our Dresden project, Tschira was interested. He would consider funding a bioinformatics group in Dresden. I informed the mayor of economic affairs in Dresden, Rolf Wolgast, and he made a list of villas near the MPI–CBG that might be suitable for a bioinformatics institute. Konrad took the list of suggestions to Tschira, who looked them over. Cunningly, Wolgast had included Lingnerschloss, a castle on the far side of the Elbe opposite the MPI–CBG, which was owned by the city. Tschira chose the castle. Why not have a bioinformatics institute in a castle? We thought the idea was brilliant. We just needed the support of the government of Saxony. Konrad and I presented the plan to Biedenkopf, who became enthusiastic about this prestigious project and supported the initiative.

Dresden had built a new air terminal, which was inaugurated in March 2000. Surprisingly, I was asked to give a speech at the inauguration. I emphasised how Dresden was becoming international again, as it had been in the times of August the Strong, the eighteenth-century elector of Saxony. The audience appreciated my appeal for internationalisation. This warmed my heart because we were dependent on strong political and public support at this stage of our project. Biedenkopf was there and told me that he would soon meet with Tschira.

We had a lot of balls in the air. Naturally, our priority was the MPI–CBG. We also had to take care of our relations with EMBL and the MPG and with the authorities and ministries in Dresden. Amazingly, our Innoregio proposal for biotechnology had reached the final round of review. We were still planning our biotechnology start-ups, and now we had Tschira's castle initiative, which further strained our limited human resources.

Remarkably, all these projects made progress. So far, we had managed to juggle them so skilfully that none of the balls had been dropped; Ivan, Marino, Tony, Wieland, Konrad, and the rest of our team caught them before they hit the ground. The collaboration worked. Our strategy of concerted action to master the challenges had proven successful. Each of us had responsibility for his own field of activity, and the steering group met regularly to discuss progress and problems. Myself, I was always trying to detect problems before they became acute. The burden of handling all contacts with politicians and authorities and of putting in appearances at receptions, parties, and inaugurations was alleviated by the fact that there were several of us. Dresden's position as the capital of Saxony also made it easier to maintain our networks.

In November 1999, our Innoregio project, BioMeT, was selected for funding. Twenty million euros; what a day of joy! Only twenty-five of the 404 applications had received funding. Our synergistic project was considered one of the most innovative. So far, however, it existed only as plans. We were rewarded for their potential. Now, it was time to start implementing them.

***

The first ELSO meeting was held in Geneva in September 2000, and, like the ECBO meeting in Heidelberg in 1995, it was cordial and welcoming, and we deemed it a success. We had attracted nearly 2,000 participants, predominantly young researchers. The quality of the programme was as high as at the ASCB congresses, and the poster sessions were lively as well. For the doctoral students and postdocs who presented their results, it was important that their presentations evoked interest. At European conferences, the poster sessions were traditionally poorly organised, and the posters did not get the attention they deserved. In our advertisements for ELSO 2000, we had emphasised that the poster sessions would be the highlight of the meeting and that those who presented their results could count on the attention of group leaders and senior scientists. Maybe this contributed to making ELSO a true working congress rather than a tourist trip for scientists paid for by others. Christian Sardet from Nice, whom I had known for many years, organised an evening session during which researchers presented videos that showed cell processes in all their dynamic beauty. He called this innovative form of presentation 'Cinema of the Cell'. Almost all participants came to the opening session. It was a terrific success, so we decided to arrange a video competition for the next congress and show the winning entries at the next Cinema of the Cell session. Our congress team had a new member, Carol Featherstone. Carol had been the Editor of Trends in Cell Biology in Cambridge and had now moved to Toulouse. She had been a postdoc at EMBL and belonged to our network. Carol was starting an online journal, The ELSO Gazette, hoping to attract advertisers.

The whole ELSO team felt enormous relief after all the stress we had been through. We decided to postpone hosting the meeting in Dresden until 2003; instead, we began preparations for ELSO 2002 in Nice.

The time for our move was approaching. On 8 December 2000, we threw a thundering farewell party at EMBL. Paul Nurse and the former director of EMBO, John Tooze, sent a video greeting from London in which, in British style, they made jokes about us all and me in particular. After that, Marino and Tony showed a video in which they were standing outside the Heidelberg railway station dressed in rags, singing and begging for their living. The skit was about how Kai had promised heaven and earth, but nothing had come of it, and the whole Dresden project had foundered. Now, they were out of work. The show continued to roaring laughter from the audience, and even Fotis was in a good mood. As a farewell present, I was given a bicycle made in East Germany before the Second World War. Konrad had found this treasure, I have no clue where. I cycled around the auditorium until I collided with a big flowerpot. The bicycle is now a museum piece, waiting in our garage in Dresden for better days. The party continued into the early hours of the morning. What nostalgia!

Our time at EMBL was over. Originally, I had expected to stay for three years, but, in the end, I was there for twenty-five. It had been a happy time on all counts. My research had flourished, and I had matured as a scientist. In my twenty-five years, Cell Biology had trained laboratory assistants, doctoral students, postdocs, and group leaders on fixed-term contracts. Now, they were active all over Europe, spreading the working methods of EMBL. Of these alumni, 130 were department chairs or professors or had leadership positions around the world. Training was one of the most important tasks of EMBL, and we had been successful. In doing so, we had built a network of EMBL alumni with the interdisciplinary collaboration skills required to solve today's complex research problems. This network was an essential consideration when we founded ELSO. Without the EMBL alumni, ELSO would never have worked, and without the experience that I had gathered at EMBL, I would never have dared take on the Dresden challenge.

***

We sold our townhouse in Heidelberg and were ready for our next adventure in Dresden. I was already sixty-two years old. It was a risky enterprise. How well would we manage the project? Would my research come to an end? Would we find new friends? Many questions were open.

We moved before Christmas 2000. The move went smoothly because the moving company did everything. They placed the furniture where we wanted it. They even hung the pictures and lamps. Still, we needed to unpack some boxes ourselves, and it took time before everything was in its place. The greatest benefit of moving is the opportunity to throw away unnecessary rubbish.

We had handled all our previous moves ourselves, improvising with friends to help us. When we moved into our Heidelberg house, we threw an unforgettable party for all those who had helped. We emptied the entire liquor store that had come with the moving van. My father Lennart was with us and in great spirits.

At Christmas, all our children got together in Dresden. Majlen, Ari and Jonne also came. The renovation of our 135-year-old villa had been successful. Admittedly, it cost twice as much as we estimated, but it was worth it. The sauna that had been built under the cellar vaults was a minor catastrophe, however. The screw-heads in the sauna bench burnt like fire. Apparently, the carpenter who built the bench had no experience with hot saunas.

The garden was a major project. It is nice to have a little green oasis in the middle of the city, but the previous occupants had not taken care of it. When we arrived, it was completely overgrown. Most gardens in the GDR were cluttered with boards and other building material useful for the flourishing barter trade. Bartering was a kind of social activity, a culture that disappeared immediately after the fall of the wall.

The MPI–CBG building was finished according to plans on 31 January 2001. The MPG had never built anything so quickly. It had only been three years since we signed our contracts. None of the building companies went bankrupt, which was unusual for such a large project. It was even built within budget. We had taken the risk of planning five floors when, in principle, only four were allowed by the building code. Mayor Just, responsible for building matters, was not known for allowing exceptions, but this time everything went well. Just was an admirer of Finnish architecture. When Heikkinen, Komonen, and I visited the Mayor, I had to act as an interpreter since neither of the architects mastered both German and English. Despite that, it was a cordial meeting, and the five floors were never mentioned.

In the early phase of planning, however, there had been a crisis. The façade was to be covered with a grid of bronze or copper. In summer, this would have a cooling function, preventing the sun from shining in through the windows whereas in winter, it would decrease heat loss. When Wieland realised that the grid would disturb the view through the windows to some extent, he wanted to change the whole façade. He would not budge. To show him how the façade would appear, the MPG built a two-storey model of it in Munich to please its German Max Planck director.

On a raw and chilly autumn day, Wieland, Tony, Marino, and I travelled to Munich to inspect the model façade, which had cost 200,000 D-Marks to build. Wieland climbed up a ladder to the second floor to look out of a window that was partly covered by the external grid. His only comment was, "All okay". But there was an unexpected benefit. The façade model had three sections with grids of bronze, copper, or aluminium. Now that we could see it for ourselves, we realised the aluminium section looked best. Neither the copper nor the bronze grids were attractive. Moreover, the aluminium grid was the cheapest by far. Thanks to Wieland's insistence, the MPG saved half a million D-Marks! Did he have a sixth sense? We returned to Heidelberg happy and contented. This was teamwork of an unusual quality.

Wieland was responsible for moving all the laboratory equipment from Heidelberg to Dresden. He did this with meticulous precision and authority. We bestowed on him the honorary title *Generalissimus Huttner*. Among the equipment were twenty -80°C freezers full of important biological samples. For safety, Wieland had ordered ten empty freezers cooled to the same temperature, which were transported in separate trucks, just in case one of the trucks with the precious samples should get into an accident. If anything happened to a full freezer, a replacement would be available immediately. The laboratory animals travelled in a separate car. The fruit flies came with us to Dresden, as did the zebrafish swimming in their aquaria. Somewhere, too, there were the nematodes that Tony was working with and salamanders that newly recruited group leader Elly Tanaka was using as an experimental model. Everything went according to plan.

At the same time, one hundred researchers, many of them with families, moved from Heidelberg to Dresden. The housing situation in Dresden was excellent. It was easy and, above all, cheap to rent a flat in a central location near the MPI–CBG. In one week, we installed all our apparatus and laboratory instruments in the new institute. Wieland had bragged that he would start doing experiments one week after moving, and he did so. Whether the experiments were just for show or were serious, I was never able to figure out.

The respected German weekly magazine *Die Zeit* sent a journalist to interview us when we moved into the new building. The reporter stayed for a whole week and also attended the inauguration party of the new institute. Marino and his gang had made a video about the move, in which they joked about us all in good old EMBL style. Wieland was given a good deal of abuse, which he took with good humour. By now, we were all well-accustomed to being laughed at.

The journalist was delighted and wrote a full page in *Die Zeit* about the MPI–CBG. He had never met Max Planck directors, who were so exuberant and unconventional. The article was the best public relations we could have asked for. Later, at a panel discussion in Berlin organised by *Die Zeit*, I met the reporter's boss. He asked me if we had given the reporter some drugs to make him so enthusiastic because he was usually so critical.

The building was a masterpiece. The architecture fulfilled all our expectations. The aesthetics and the choice of materials were outstanding. Heikkinen and Komonen were masters in creating an inspiring atmosphere with minimal means. The building was functional and facilitated communication, just as we had hoped. We had an argument with our own administrative department, which wanted to exchange the glass doors and walls of their offices for opaque materials. They were disturbed by the prospect of an open view into their offices. But we insisted on the glass to encourage transparency and service. We wanted to emphasise that the role of the administration was to support us. That was the end of the matter; the administration soon reconciled itself to the architecture, which did indeed support our goals and promote an atmosphere of openness and togetherness.

Marino was not only our show master. As an Italian, he also took responsibility for the canteen and the cafeteria. It was essential that everyone who worked at the institute could have lunch together in the canteen. Marino had struggled to find a chef who could satisfy our requirements for international and healthy cuisine. This was not something that Saxony was known for. He also spent a great deal of time ensuring that the coffee in the cafeteria met Italian standards. The MPG had installed a German coffee machine, but it did not serve espressos of the quality that Marino wanted. So he bought an Italian machine. Now, we could choose German coffee or Italian espresso.

We insisted on details and perfection. A little luxury is needed even at a research institute.

Tony was in charge of procuring all the scientific instruments at the MPI–CBG and was also responsible for information technology. Two young Americans who were summer interns in his group at EMBL were both real wizards in this field. They stayed in Germany and came to Dresden, where they founded the start-up Scionics, which became the main provider of information technology services to us and many others. It was such a pleasure to follow the development of these and other youngsters who joined our MPI–CBG team. The dominance of youth in our institute gave us all such a boost and pulled the whole enterprise forward. The collective enthusiasm was contagious.

***

Wieland had agreed to set up a doctoral training programme modelled on that at EMBL. For me, the recruitment of doctoral students was the greatest question mark that we faced. Would we be able to build an institution with as many as twenty-five functioning research groups? How could we attract young, talented scientists to Dresden of all places? We would have to recruit from outside the city, as modern biology at TU Dresden was not yet developed enough to train doctoral students. Even at EMBL, we had problems starting our graduate programme. In Dresden, things might be significantly worse. I thought we would have to organise lectures, symposia, and other events in Warsaw, Prague, Budapest, and Bratislava to recruit researchers.

We announced that the MPI–CBG would start its doctoral programme in 2001. Max Planck President Markl had launched a new initiative: international MPG doctoral schools to be organised in collaboration with neighbouring universities. Wieland proposed to set up a doctoral school with TU Dresden, which was approved.

When the deadline for applications to the school passed, we could not believe our eyes: our programme had received lots of applications. The quality of the candidates was higher than we had dared hope for. Times had changed, and the internet now facilitated our task. Many applicants were attracted by the new combination of cell biology, developmental biology, and physics. Another factor was the reputation of the team of leaders, whose achievements could be found easily on the web. The focus on the physics of life was a particular attraction. We invited thirty top candidates to spend a week at the MPI–CBG. During that week, all five directors and twelve group leaders searched for doctoral students, and the thirty candidates looked for appealing group leaders. The process was inspiring but intense: the relations formed had to last for four years of collaborative work. Joe and Karla had not yet moved from Seattle, but they came to Dresden to participate in the selection process. Most important of all was that the students whom we selected accepted our offers and came to Dresden. My big question mark had become an exclamation mark; what a relief!

It was time to consolidate our scientific programme at the MPI–CBG to develop a synergistic project that would give substance to our slogan 'How cells form tissues'. That required intense discussions about the plans of each research group and whether their goals were achievable. We gathered for a retreat at a hotel in Erzgebirge in the mountains of Saxony. Here, all the directors and group leaders presented their plans, and each presentation was followed by lively discussions. In the end, we had a general debate about everything that had been presented. There was no general consensus, but glimmers of collaborative projects could be perceived. What more could one hope for?

Ivan Baines had begun to establish the infrastructure for our biotech projects at the MPI–CBG. Also, he was involved in founding Biopolis Consultants, which would energise the development of biotechnology in Dresden. Ivan added enormous value to our team of Directors. The sextet was extraordinary; it is amazing what energy is generated when teamwork functions.

Our biotechnology project in the city was branded Biopolis Dresden. We were not the first with this idea: Biopolis Singapore existed before us. Singapore was a giant, and Dresden was a dwarf. Nonetheless, we could advertise ourselves in the media under the logo of Biopolis Dresden. When journalists asked us if our goals were unrealistically ambitious, we admitted that this was probably the case but that did not prevent us from trying. Also, I referred to the Nokia miracle in Finland. These tactics made us standard-bearers for the future of high-tech in Dresden. We wished to establish biotechnology as a complement to the microchip industry, where Dresden had already established itself as a European leader.

Our Innoregio project, BioMeT, needed to get off the ground. At one stage, it threatened to slip out of our hands. GWT, the organisation for commercial projects at TU Dresden, quite shamelessly had begun to distribute support to their insider circle before we discovered what they were doing. I had to take a strong line and called the responsible people at GWT to a meeting. I threatened to go to the media unless they listened to reason and introduced a structure with expert evaluation for BioMeT. Only realistic projects should receive funding. Led by Ivan, we convened a committee to assess and select commercial projects that qualified for BioMeT support. Anyone with a suitable project in Dresden could apply.

It is incredibly difficult to distribute public money for commercial research and development projects in an expert manner. The problem is the same everywhere in the world, and especially difficult in Dresden where this expertise was totally missing. Luckily, we managed to help found Biopolis Consultants with Ivan and two experts Ann de Beuckelaer from Belgium and Mark Hentz from Hamburg, and they received a three-year grant from the government to develop both Biopolis Dresden and biotechnology in Saxony. Without this support, our Biopolis Dresden project would undoubtedly have withered.

***

Klaus Tschira's bioinformatics project, now known as Bioparc, was materialising in a surprising form. Tschira himself had single-handedly sketched a building shaped like a DNA helix for the park of the castle, as a companion to the castle itself. He was very proud of his creation and wanted to have it built. I hoped to engage Heikkinen and Komonen to give architectural shape to the helix idea, but Tschira already had a trusted architect who administered all the building projects for the Tschira Foundation.

Who would fund this huge Bioparc project? That was now a crucial question. Tschira did not want to do it alone. Biedenkopf had committed to Bioparc, and we hoped for yet another miracle. Bioparc had become a visionary project combining baroque with modern technology for Biopolis Dresden. That combination fitted Dresden perfectly. The challenge was to connect a glorious past to the future. Bioparc would become a symbol for this effort.

***

My research group needed attention, but I could squeeze so little time out of my packed schedule. My friend Herbert Jäckle noted that when he became Vice President of the MPG, he realised during his morning shower that his brain was already filled with administrative problems, and that this was bad news for his research. I felt the same. I did not have enough strength for all the projects we were working on. When you leave your lab adrift, the research will move away from its goals, often according to the law of least resistance. If you don't watch out, the whole project you have planned may fragment and fall apart. Therefore, I had to be there to coax, persuade, and help when my researchers got stuck and, at the same time, keep them on track. We can never know in advance what strategy will bring the correct solution to the puzzle we are trying to solve. Sometimes, I had to accept that moving sideways might bring new light to a research problem. I should not stubbornly insist on a previously agreed strategy. I had to be flexible. Most important of all, my physical presence was needed now and then.

To untie the knots in our research, scientists need contact with other scientists, not only with other members of the group. When experiments fail, researchers instinctively want to disappear from work without being seen. Yet, meeting a colleague just by chance may inspire you with an idea to solve your problem. Sometimes, it may be enough to describe your difficulties to someone who understands what you are doing for you to find a simple solution. When thoughts go round and round in your head, you need some outside push to break out from the paralysing deadlock.

When we planned the routes of circulation within the institute, we decided to close all the exits except the main exit from the central atrium of the building to increase traffic through that space. Thus, the main exit also became the main entrance of the MPI–CBG. The atrium was dominated by a spiral staircase, which was one of the most impressive features of the building. This staircase became a communication portal. The architecture served its purpose! Many preferred to walk rather than use the elevators. Ideas were born, and problems were solved by chance encounters on the way up or down the stairs.

During this exacting time, Carola was my angel and even worked part-time in the lab. She had decided not to open a new dental office in Dresden. The investment would be too heavy for the few years she wanted to work before retiring. After the move, we had problems with the MDCK cells. They would not grow like they did in Heidelberg.

Cell culture had been Hilkka's responsibility, but she retired before we moved to Dresden. Carola managed to find out what was wrong. Our new laboratory assistant had changed one detail in the protocol without consulting me. When we reverted to the old protocol, the cells grew perfectly again.

Carola gave me strength through her understanding and pragmatic support. I had stopped smoking my cigarillos. One time, I secretly smoked in the garage and was caught out. After that, I never smoked again. Did I get fewer ideas than before? It was hard to tell; scientific ideas were becoming rarer anyway due to competition with all the other projects that troubled my mind.

***

We devoted a lot of time to finding an excellent bioinformatician for Bioparc. At last, we had a perfect candidate: Chris Sander. He and I had been colleagues at EMBL, and now he was professor at Harvard. I met him while in Boston for a Biogen meeting and managed to interest him in Bioparc.

But sadly, Bioparc went nowhere. The negotiations rolled back and forth. Each time I thought we had made progress, a new problem cropped up. One of the three leaders – Tschira, Sander, or Biedenkopf – always came up with objections that delayed the project in some way. An investment of this magnitude was not easy to achieve. To make matters worse, a group of conservatives in the city started opposing the project. There was a stand in the city centre, distributing leaflets against the DNA helix building. They did not want a modern building in the castle park.

Dresden is indeed conservative, for better or worse. The stubborn conservatism of the inhabitants had helped them protect the bombed-out city centre from being rebuilt with dreary GDR architecture. This meant that when the wall fell in 1989, the heaps of rubble were still there, and the destroyed buildings, like the *Frauenkirche*, could be rebuilt to their former glory. This concern for the past, however, made looking forwards more difficult.

Another issue was that many of the most active inhabitants of Dresden no longer lived there. They had had enough of the repression and the problems of the GDR regime and had moved away as soon as they could. This exodus marked all of East Germany after the reunification. Without the wave of immigrants from the West, Dresden would not have flourished. Many of those who were now engaged in rebuilding Dresden were newcomers.

This was a recipe for conflict. Hermann Krüger, retired chief of the Dresden office for historical preservation, rang our doorbell one weekend. He came to complain about the project in the castle park. He started by remonstrating about everyone who had moved to Dresden after the fall of the wall, arguing that Saxony did not need these newcomers. He even mentioned Kurt Biedenkopf as one of those who were not welcome. Then I interrupted him and asked why he had come to us, "We are newcomers, too." His answer was typical, "But Professor Simons, you come from Finland." Many East Germans by tradition felt an affinity for the Finns. Then he started to criticise the Bioparc project, and I just listened to him talk. Trying to change his mind would have been hopeless.

Unfortunately, the anti-Bioparc movement attracted more and more followers. It was still a small minority in Dresden, but Klaus Tschira was unhappy with the commotion. The movement also spread to TU Dresden. We had persuaded Chris Sander to apply for a chair in bioinformatics, which had been created especially for him. After Chris gave his lecture, he was interviewed by the appointment committee, of which I was a member. A major issue for some of the professors was whether Sander would be willing to teach basic courses in bioinformatics. The discussion in the committee after the interview was hair-raising and embarrassing. Chris, an internationally renowned bioinformatician, who was prepared to move from Harvard to Dresden, was treated in such a derogatory manner. In the end, the committee unanimously supported Sander's appointment. Two weeks later, however, a delegation from the Faculty of Informatics came by with the message that they opposed the appointment. They had blocked Sander. This was unbelievable and parochial beyond belief. What did the informaticians in Dresden fear? That their microcosm would be turned upside down by confrontation with a world-class researcher?

Now, even Tschira began to falter, and he put the project on ice. Chris had no option but to withdraw. Building Biopolis Dresden was not an easy challenge, but we just had to grit our teeth. After all, Bioparc was just one of Biopolis Dresden's projects.

**

Finally, we found success again. The government of Saxony granted us 100 million euros to implement the bioengineering project. That meant a building with space for six new professors in a new Bioengineering Institute, we called Biotec, and additional premises for commercial biotechnology. What we had wished for was coming to fruition: academic research and biotech start-ups under the same roof, a centre for bioinnovation in Dresden. At last, I could tell Francis Stewart that all was ready. His confidence in us had been justified. A weight fell from my shoulders.

Our institute, the MPI–CBG, was officially inaugurated in March 2002. It was a grand event with the German Chancellor, Gerhard Schröder, as guest of honour. The Max Planck leadership was present, led by Hubert Markl. The MPG was beginning to realise that we were perhaps an asset to the entire organisation after all. It certainly helped that we had managed to get our new institute and Biopolis Dresden into the national newspapers. These were still early days after reunification, and the media were eager to report on new developments with positive potential.

Markl was worried that the budget of the MPG could not sustain its expansion in the East. In a period of about ten years after the wall came down, the MPG had built eighteen new institutes ranging from anthropology, ethnology, history, and psychology to physics, chemistry, and biology. This investment reshaped the German research landscape.

With Markl, we pondered whether we might ask Chancellor Schröder for an increased budget for the MPG. Not easy, of course. Everyone wanted more, and he received many such requests. I suggested that I might tell him about Finland's dramatically increased investment in research after the collapse of Soviet trade in the 1990s. The Finnish prime minister at that time, Paavo Lipponen, who was responsible for this policy, was a Social Democrat like Schröder. Counterintuitively, he had increased the research budget when unemployment was at a record high, and this had a positive effect on managing the crisis. This investment also put wind in the sails of Nokia.

Schröder came earlier to the MPI–CBG inauguration than we had planned. I had to entertain him alone until Markl and his team showed up. I found the Chancellor friendly and easy-going, and we had a long discussion before Markl arrived. At the small reception before the inauguration ceremony, when I told him about the unusual research policies of Finland in the 1990s, his reaction was less amiable and abrupt. "*Was wollen Sie von mir, Herr Simons?*" ("What do you want from me, Mr Simons?"), he asked with irritation. I did not know what to answer. Fortunately, Markl saved me with small talk.

After the reception, we all gathered in our large, new lecture hall. Schröder gave his speech without a script and included a long passage about Lipponen and the good relations between Germany and Finland.

His speech was wonderful. Had our dissonant exchange had a positive effect after all?

Finally, Tschira withdrew from the Bioparc project. It was a pity but not unexpected. The protests of the Dresden inhabitants had cooled his interest, but there were other reasons, too. Afterwards, he seemed to feel bad about it and compensated with a major donation to ELSO. Also, Konrad persuaded him to fund a group leader position in bioinformatics at the new Bioengineering Institute. Thereafter, Tschira remained supportive of us, which was a great relief.

Still, it was disappointing to give up the Bioparc project after all the excitement. For some reason unfathomable to me now, I started thinking of a new idea. My plan was that Saxony should forge an alliance with Estonia, Finland, and Flanders to set up common goals in high-tech research. One important argument was that such an alliance could be a model project for EU funding. All four states were dynamic and ready to invest in new ideas. I guess I wanted to confuse those who had followed the debacle of the Tschira project by proposing a bold new initiative.

I happened to be invited to give the first Tanner lecture in Helsinki. Tanner was a Finnish social democrat politician. I launched the idea in my lecture and discussed it with social democrat politicians during the dinner that followed. My lecture was titled '*Uljas uusi biologia*' ('Brave New Biology'). The theme was how biology and biotechnology would increasingly dominate the twenty-first century. The alliance project might help prepare the countries for this development and open new commercial opportunities. Also, I presented my ideas about molecular bioengineering with bionanotechnology as a platform for engineering solutions for the future.

I had been invited by Mart Saarma, who was head of the Institute of Biotechnology at the University of Helsinki. Mart came from Estonia and promised to promote the alliance project there. Around the same time, I was also invited to the University of Leuven in Flanders, which conferred on me an honorary doctorate. During the festivities, I took the opportunity to present my project to the research minister of Flanders.

At home in Dresden, I continued to promote the four-state alliance project and hoped that some progressive politician would take up the idea and carry it forward. Nothing happened, of course. How stupid could I be? We had no team for this project. My escapade in Moscow should have taught me that nothing happens by itself. Promoting a project like this one requires intense, goal-oriented action over several years. We had no time for that. Later, acquaintances in the ministries of Saxony told me that they had my memorandum about the alliance lying somewhere in their office and thought the project was potentially interesting. But they themselves never thought of doing anything to promote it.

Carola consoled me and asked me to be reasonable, "Focus on your main activities! You'll achieve nothing if you run around like a headless chicken." She was right, as usual. Carola is indispensable.

***

It was time for ELSO 2002 in Nice. Carol, Ingeborg, Konrad, and I had worked hard to plan everything. At one stage, we had no money left, and neither Konrad nor I had time to look for sponsors. We were fully occupied with the MPI–CBG project and all its facets. Carola had sold her dental office in Heidelberg. The sum should have been her pension, but I managed to persuade her to lend the money to ELSO. She did, and it saved us.

Now we had to secure participants for the Nice meeting. I e-mailed our entire network almost every day to remind people about the congress. We relied mainly on EMBL researchers and alumni. My emergency calls became legendary and were effective. We had one thousand poster abstracts, which was a record. Maybe the whole congress would be equally successful?

It was. Everything exceeded expectations. We had learnt how to organise congresses for 2,000 participants. Carol had founded an ELSO committee preparing initiatives relating to research careers. Among the group's goals were improving the position of women in the research community and reforming the European research career ladder. We wanted to use ELSO as a platform to propose improvements to both the EU and the national governments. ELSO was determined to realise the Euroexcellence project for young scientists proposed by the EU committee that I had chaired.

After Nice, we again heaved a sigh of relief, but there was no time to rest on our laurels. Preparations for the next congress, ELSO 2003, in Dresden, would have to start immediately. Our work rolled on inexorably. We had started a project that we thought Europe and Dresden needed. Now, we had to see it through.

***

Together with Ari and Majlen, we bought a piece of land with an old wooden villa on it on the shore of Mölandet, an island off the Nordsjö neighbourhood of Helsinki. It abutted the lot with a summer cottage on it that my father Lennart had bought before he died. Majlen and Ari renovated the villa, and Carola and I built a new summer house. My brother Tom and his family took over Lennart's house. All three Simons siblings now had summer homes next to one another. Mölandet became a pivotal point for the whole family when we met in the summer. We wanted to keep roots in our homeland.

After the ELSO congress in Nice, we had a lovely summer on Mölandet. A high-pressure weather system parked over Finland, and the sun shone every day. The algae in the sea were our only problem, but not even that pest lasted for long. New regulations limiting run-off from the land, I hope, will stop or at least decrease the eutrophication of the Baltic Sea and the Gulf of Finland.

A few days before the holidays ended, we received dramatic news from Dresden. It had been raining cats and dogs for two days. The water level in the river Elbe was rising day by day. When we returned to Dresden, it was nearly ten meters higher than normal. The torrential rain led to terrible flooding.

The situation looked threatening for our institute on the shore of the Elbe. Fortunately, in the planning stage, Wieland had raised the question of what would happen when the Elbe flooded. We had experience from Heidelberg, where the river Neckar flooded sometimes. The most destructive flood in Dresden previously had occurred in 1845, when the Elbe had risen by almost nine meters. The MPG had agreed to build the MPI–CBG to withstand a flood of the same magnitude.

Luckily, the water did not rise high enough to flood the institute. The building was constructed like a bathtub, which prevented groundwater from invading the building from below. All our expensive equipment in the basement was well protected. Many other buildings in Dresden were not so fortunate. The pressure of the groundwater forced water up into them. In the *Landtag* (state parliament) of Saxony, the basement garage had to be filled with water so that the building would not float away. Our five-story building was heavy enough to stay in place.

Our villa was safe as well. It was not far from the Elbe, and the 1845 flood had probably been a fresh memory when it was built in 1865. By contrast, Marino's house was invaded by rats that crept up from the drain in his cellar. He had to kill them with an air rifle.

Most remarkably, the railway station, which was not even close to the Elbe, was flooded. It happened so suddenly that passengers who were waiting for a train had to run to escape the streaming water. A tributary of the Elbe had found its old riverbed, which ran right through the station. The mass of water rushing down from Erzgebirge refused to follow the drain around the station building. No one remembered that the tributary once followed this course.

The flood evoked immense solidarity. Hundreds of young and old volunteers came to Dresden to build barriers of sandbags. When the water masses subsided, they helped to clear away the mud that covered the city. Others offered food or free lodging to the volunteers. The inherent capacity for mutual aid was expressed everywhere. We are not so selfish as is generally assumed. It felt good to experience this wave of altruism.

Chancellor Schröder himself came in rubber boots and helped for a few hours. This was shortly before the parliamentary elections. The pictures of him in the media certainly won him a few extra votes to ensure his re-election. For Dresden, the most important outcome was that the government appropriated a sum of the order of billions of euros for restoring the city after the flood. The money gave the economy of the city a much-needed lift.

Slowly, life in Dresden went back to normal. I was devoting more time to the lab to ensure that our research advanced. My lipid rafts, which were at the centre of our efforts, had been exposed to savage criticism. We produced new data, and results from other laboratories also supported the concept. There were also more negative reports, however. Some of the methods used for studying rafts in membranes were, rightfully, questioned. Our methodology was indeed still inadequate. The researchers at the LMB in Cambridge were especially critical. Sean Munro published a review entitled 'Lipid rafts: elusive or illusive', with an emphasis on 'illusive.' Rafts had their ups and downs, just like all my projects.

When new findings and hypotheses open new perspectives, it is normal that there is a counter-reaction. That is how science works. Advances are tested critically, and it takes some time before the field accepts or rejects a new hypothesis. So far, I was not worried; all the negative findings could be explained. We just had to struggle on.

My rafts had invigorated membrane research and drawn many biophysicists into the field. It boded well for the future. Most importantly, I was able to attract high-quality doctoral students and postdocs who enthusiastically committed themselves to our membrane research.

Like all researchers, though, we had difficulties getting our results published. For us, there was more resistance than usual because of the controversies surrounding our concept. A single negative review by a referee was sufficient to get the article rejected. This happened quite often, but as a member of the National Academy of Sciences of the USA, I had the privilege of publishing in the academy's journal, Proceedings of the National Academy of Sciences USA, where members could choose the referees themselves. This privilege kept us afloat.

In science, we are evaluated constantly. Every internationally recognised scientist is bombarded with a stream of requests to act as a referee, assessing manuscripts and grant applications, evaluating applicants for professorships, and examining doctoral theses. These tasks can be carried out at one's own desk. But, on top of that, there are trips to evaluate research institutions and larger research projects.

Ivan and I were tasked with evaluating biomedical research in the whole of Norway. We were confined to a motel for a whole week, sandwiched between a gas station and a McDonald's near Oslo airport. Professors and group leaders filed past us in an uninterrupted stream, which petered out only when the week was over. It became obvious that the research was not up to international standards. The groups were small, often comprising only two people. Amazingly, the greatest problem was funding. Despite its lucrative oil profits, Norway did not invest much in research. Maybe the reason was that research in Norway was long led by what was called the Ministry of Ecclesiastical Affairs, literally, the Church Ministry. The purpose of our evaluation was to give this ministry general advice on how the molecular biosciences in Norway could be supported. Unfortunately, our proposals for reform fell upon deaf ears.

I had a similar experience earlier. I was a member of a committee preparing an initiative to build a global centre for marine biology in Bergen, Norway. The idea was to create something big. Marine biology is full of exciting potential but is globally under-researched. We were all enthusiastic about the project. This was a glaring gap that Norway was well-positioned to fill. The country has a big aquaculture industry, which would profit directly from such an investment. Naturally, basic research would play a central role in the research programme, however, it would also include applied research on prevention of fish diseases and methods for protecting the environment from the negative effects of fish farming. Aquaculture was growing all around the world. We travelled home happy to have been involved in supporting such a fantastic project. The sea is full of life, poorly studied until now.

But what happened? A battle over resources ensued. The other university towns in Norway also wanted marine biology institutes. Finally, the government decided to forego the plans to invest heavily in marine research. In familiar Norwegian style, Bergen had to content itself with a modest institute, which was founded in 1997. It never became the centre that Norway and the world needed.

The burden of evaluations was much lighter for previous generations. When my father was asked to evaluate the applicants for a professorship in Stockholm, he was excused from lecturing for a whole term in order to engage in the task. He also told me that Max

Planck was once asked to act as expert for the physics chair at the University of Helsinki. He answered with a postcard saying, "Take Jarl Wasastjerna, I know him." Wasastjerna got the position.

The MPG evaluates its institutes every three years. In 2003, the MPI–CBG was evaluated by an international group of experts tasked with analysing how the institute had fared internationally. We were well prepared. Our presentations were first-rate, and we made no blunders. The committee was positive. We could heave a sigh of relief. The first period could not have gone much better. The outcome of this first evaluation raised spirits at the institute, which was needed after such an exhausting time.

It was a great bonus that we had managed to build a truly international institute. The MPI–CBG has attracted young researchers from nearly thirty countries! At the next evaluation, five years later, we counted forty-three countries. Nobody, not even I, would have thought this was possible in Dresden. I had underestimated the power of the internet to disseminate information about us, which attracted researchers from near and far. Before us, there were hardly any foreigners in Dresden. We had created an unusual ambience. It was not just the quality of the research or even the atmosphere of the institute. The quality of life in Dresden was an important factor, too. Many of the young people working at our institute lived in Neustadt, on the other side of the Elbe, but within biking distance. When we first came to Dresden, trees were growing out of the windows of the decayed houses in Neustadt. Now, these nineteenth-century houses had been renovated, but not so heavily that rents skyrocketed. In the West, rents in cities with top-level research institutions had reached the point where students and postdocs struggled to make ends meet. Neustadt offered quality of life. The district was full of inexpensive little bistros and bars that simmered with life at the weekends.

***

We worked hard to build local momentum for our projects. The MPI–CBG featured in regional and national newspapers and magazines, and *Der Spiegel* wrote a lavish article with the headline 'The ballet of life'. We wrote about Biopolis Dresden in the international professional media, and ELSO spread our message in Europe.

One problem was that the articles often focused on me as the figurehead of the project. That was not what we wanted, but it is how the media work. It was something we couldn't do much about. According to the popular daily newspaper *Bild*, I was the third-best-known person in Dresden in 2002. After that, fortunately, I dropped out of the list.

The publicity campaigns made it much easier to find the contacts and support that we needed for our projects. Bioparc had misfired, but even that failure gave us extra points since the blame was put on conservatives in Dresden who had organised the resistance. The arguments of the opponents won no support outside Dresden. Instead, the failure evoked sympathy for our efforts.

Thanks to the publicity, we were also recognised by personalities on the cultural scene in Dresden. We were visited by both the Intendant of the Semper Opera, Gerd Ueckert, and the General Director of the State Art Museums of Saxony, Martin Roth. They found it exciting that top-level researchers in biology had settled in Dresden. I suppose they had their feelers out to see whether we might have something to offer to them – just like us, they were in pursuit of new vistas. A new century needed novel impulses.

Biology was moving away from the simplifications of reductionism. But could we move further and devise projects where art and science could join forces? Would it be possible to complement the ever-changing scientific knowledge about ourselves and our planet with the quest for moral and ethical norms for humankind and our fragmented society? Utopian thoughts passed through my dreaming mind.

In a society populated by narrow-minded specialists, it is difficult to build bridges. Individualism is still too dominant. The commercial pressures of our global economy are overwhelming. We need room to breathe.

Marino wanted to celebrate the architecture of our MPI–CBG building and the work of Swiss artist George Steinmann that was displayed in our atrium. Marino edited a book called 'Gentle Bridges: Architecture, Art and Science'. Steinmann had designed large photographic panels picturing lichens, which are stable, symbiotic associations of a fungus with an alga or cyanobacterium. His art acknowledged our pursuit of a synergistic environment where working together added value to our quest for new knowledge. Bridges between science and art need to be nurtured.

In a modest step to bring art and science together, the MPI–CBG sponsored an annual art award. The idea was partly inspired by a lecture by the renowned cell biologist Jeff Schatz from Basel, Switzerland. A conservative Swiss group had started an initiative to prohibit research using DNA technology in the country. Schatz recounted how the crisis had arisen. Such prohibition would be disastrous not only for academic research but also for the pharmaceutical industry, one of the pillars of the Swiss economy. The initiative was voted on in a referendum, and the catastrophe was narrowly avoided. Schatz noted that housewives and artists were the two groups that had most strongly supported the proposal to forbid DNA research. For me, this was an indication of how poorly we researchers had managed to inform the public about the significance of our work. The fact that all vaccines against the SARS-CoV-2 virus today are products of DNA technology is just one example of how important this research is.

The goal of our art award was to stimulate contacts between cell biologists and artists. The professors at the Dresden Art School would select one of their Masters students for the award of 5,000 euros. The awardee would spend three months, part-time, with a research group at the MPI–CBG to create a piece of art. The artist would gain insight into the research being performed, and the researchers would learn about how art is created. We hoped that this bridging between the art and science would increase the understanding of our activities. Cell biology is full of fantastic images from the nanoworld of the cell. We see so many fascinating pictures and videos of this microcosm. Like artists, we explore the unknown. Why not be sisters and brothers in our pursuits?

The works of art created by our awardees were all worth their prizes, and many teachers from the art school were present at the annual award ceremonies. The press was also there, as we had hoped, so the award received publicity in Dresden. The pieces of art were highly varied. I remember a nano-slipper for a *Paramecium* (*Pantoffeltierchen* or 'slipper animal' in German; a slipper-shaped single-cell microorganism) created by Grit Ruhland. The slipper was so small that it could be seen only through a magnifying glass. It was later shown in an exhibition on nanoart in Bergamo in the north of Italy, which I helped organise.

I had been invited to a science festival in Bergamo, where I gave a talk to the general public. I imagined that the audience would consist of students, researchers, and interested citizens, but when I entered the big lecture hall, I was shocked to see that it was filled with elderly people and noisy families with children. How should I manage this? I had no time to change the slides I intended to show, and I knew no Italian. My talk was to be simultaneously interpreted. This spelled disaster.

I had to think quickly. I would have to entertain my audience. I would have to play-act and do my utmost to simplify my message. The remarkable thing was that the audience sat listening in silence; even the children appeared attentive. When I finished, I was surprised by a long applause. They even stamped their feet – I was in Italy!

My show must not have been too bad because I was later contacted by the festival organiser, Stefano Raimondi, to help organise the nanoart exhibition in Bergamo. I wrote an article, 'Aesthetics of Survival', to introduce the exhibition and would like to cite a passage from it here. "We have to tease the nanotechnological machineries out of the appropriate cellular sources and adapt them for the future technology base. Future engineers should learn how to use biology as a source of inspiration. Nature is the mother of inventions! But to achieve these goals, we must mobilise the brightest and the most talented to join in this race against time. This is also why art – this exhibition of nanoart – is important. Artists are often spearheading the changes that pave the way for the future, and this is exactly what this exhibition is about. This art has a hidden message. By seeing the invisible, we are preparing our minds for the change. Our future is in the invisible."

Another work created by one of the MPI–CBG's annual art awardees Moritz Liebig, was a copy of the sculpture of Max Planck's head that is exhibited in every Max Planck institute. Moritz was spending his three months in Suzanne Eaton's research group, which worked with fruit flies. He crafted the copy of Max Planck from the special wax that is the favourite food of the flies. The head was placed in a transparent plastic cube populated by a swarm of fruit flies. The audience at the award ceremony watched the video and witnessed how Max Planck was slowly devoured and disappeared. In real-time, it took a week before he was gone. This award caused a stir, and we got an angry call from the MPG headquarters in Munich accusing us of going too far. We did not think so at all. Sometimes art can be very funny.

It may have been thanks to our cultural activities that I was invited by Bernhard von Löffelholz, chairman of the Culture Senate of Saxony and the Culture Council of Dresden, to join a project to resurrect the *Festspielhaus* in Hellerau, a suburb of Dresden. The project would restore the buildings and activities of this festival hall to their former glory. Hellerau Garden City and its furniture workshops were a forerunner of the Bauhaus movement in Weimar. The intention was to make Hellerau a centre of modern culture in Dresden. In the 1910s, Hellerau had pioneered modern dance. Löffelholz wanted to engage the famous choreographer William Forsythe for Hellerau. The project needed political support, and Löffelholz thought that my relations with the state government might help.

The project succeeded. Hellerau not only became a centre for modern dance but developed into a European art centre encompassing architecture, music, theatre, and dance. In Dresden, Hellerau was once again the address for new cultural trends. Our young researchers, of whom 40% were foreigners, also went to see the cultural programmes in Hellerau, especially the dance, for which no knowledge of German language was needed.

I met William Forsythe and learned how he worked as a choreographer. In a lecture he gave, I discovered that his way of creating new choreography resembled my research strategy. When Forsythe had an idea for a dance, he first described to his dancers in broad strokes how he imagined the dance. Then, he guided the group in shaping and practicing the movements that would be introduced into the ballet. Forsythe was continuously looking for new movements and pushing limits to discover what the group could achieve physically. In the end, this all had to be formed into creative choreography.

That is more or less how I worked, too. My research group broadly knew what our goal was, but it was not clear how we could attain it. My task was to chisel out subprojects with each group member that would contribute to revealing features of the whole goal that we were pursuing. This common quest for solutions meant that both the whole group and each individual researcher could identify with the project and, above all, find a way forward each time a new obstacle arose. Constant adjustments were necessary. Forsythe's work and my own were equally difficult in their own ways. We both pushed our groups to achieve more than they had hitherto.

In his book 'Authority and Freedom', Jed Perl writes that "writers, composers, choreographers, painters, and sculptors all aim to establish a solid foundation for their imaginative flights. The argument can be made that the wilder the flights, the more solid the foundation needs to be. Authority and freedom are lifeblood of the arts. Authority is the ordering impulse. Freedom is the love of experiment and play." The same principles apply to scientists. We need to be trained to master the foundation of our fields, which is ordered so we solidly understand what we already know. This is the basis for our flights into the unknown. To make new discoveries, we experiment and play.

Forsythe fell victim to burnout and had to leave his troupe. Keeping one's position in the elite is demanding. Sometimes, I wondered how long I could keep up the pace. In May 2003, I turned sixty-five, and sixty-five people came to my birthday party in Tomar in Portugal. We combined it with a cell biology conference. Ralph Gräsbeck, Leevi Kääriäinen, and many other colleagues, former doctoral students, and postdocs were there. The party was a fine mix of science, nostalgia, and camaraderie. The social programme was in the well-tested EMBL style with music and humorous speeches. Ira Mellman had composed a song for me, as had Wieland. Both accompanied themselves on the piano, and everyone sang along. What is there to add? It was fantastic!

My hourglass was running out of sand. I had three years left before I must retire. I had to define how I wanted to spend my remaining time in research. The MPG had agreed that I could continue with a smaller group after sixty-eight and I had privately set my limit at seventy-five.

I avoided speaking openly about my age; I did not want to be reminded that my life as an active researcher was nearing its end. I was a victim of a common syndrome among elderly, active persons. To keep fit, I adopted a routine in the 1990s, a twelve-minute gymnastics routine ending in forty push-ups. I completed this ritual every morning without exception. Always the same mantra, "Keep working!" Also, I walked to work almost every day. That put 8,000 steps on my iPhone and was an excellent way of collecting my thoughts for the day.

Now and then, I played football with my colleagues. Football became a regular weekly event in 2011, thanks to Tony. His sons went to the International School of Dresden, and Tony played football in the school gym with the other dads. The leader of the football dads was a professor of communication at TU Dresden who also chaired the board of the International School. As an outsider with no children at the school, I had to strike a deal with him to be accepted into the group: I agreed to become Chair of the Fundraising Association. Our games on Thursday evenings are still the highlight of my week. It's not all about football, however. Afterwards, we relax in a pub, where we discuss our lives and our planet. My football buddies are younger than I am. What a privilege for me! They have various professions and have travelled all over the world. Our discussions are lively and, at the same time, cordial; football unites us. We also have a social life together, including our spouses.

***

In 2003, we organised the ELSO Congress in Dresden for the first time. It was held in the Palace of Culture in the centre of the city. Our team of Carol, Ingeborg, and Konrad had done their work, and we were well prepared, as always. A successful ELSO 2003 would support both the MPI–CBG and our Biopolis Dresden project. We wanted to attract as many researchers as possible to Dresden so that they could see for themselves how beautiful and interesting the city had become. The restoration work was not yet complete, but the city's original glory was discernible again.

We had persuaded Sydney Brenner to give the opening lecture, and he did that brilliantly before a packed house. No other molecular biologist could match his eloquence. We had a dinner party at our home for congress guests. Carola sat next to Sydney and discovered that he was a fan of the Finnish comic strip Moomin and had read the Moomin books to his grandchildren. Carola knew her Moomin by heart, and they showed off by quoting the illustrator and author Tove Jansson for us all.

The congress programme worked well this time, too. The Cinema of the Cell session was again a smashing success. The video clips of dynamic processes in the cell were an aesthetic pleasure. Also, the poster sessions attracted many visitors. Crowds of young researchers formed around the posters, discussing experiments and data, and senior researchers joined in, too. This was exactly how we hoped ELSO would work.

The conference culminated with a party at the Ministry of Finance. The participants danced until we were thrown out. We were so relieved that, once again, the congress had been a success. This was balm to the soul after all the intense work. It is not easy to organise a congress with almost 2,000 participants on a minimal budget. The personal style of our small team made all the difference. The success of ELSO 2003 gave us strength to start work on ELSO 2004 in Nice.

ELSO 2003 also put wind in the sails of our projects in Dresden. The Medical Faculty, which was the MPI–CBG's neighbour, celebrated its tenth anniversary. Before the fall of the wall, TU Dresden had had no medical faculty. Everything was built from scratch. For us, it was a great advantage that the structures were not fixed. The newly appointed professors were looking for partners for collaborations, which opened doors that would normally have remained closed. The medical professors not only wanted to build up excellent clinical activity but also wanted to start their own research. This was a good basis for our efforts to realise our outreach goals in Dresden. We planned joint projects, which were taking off. The most important of them was a collaboration between regenerative medicine and stem cell research. Our MPI–CBG salamanders could regenerate lost body parts; maybe one day, science could learn how humans might do the same. We wanted to translate the basic research done at the MPI–CBG into clinical practice, and this goal energised our cooperation.

The 100 million euros granted by the State Government were now providing the means to build and establish a Bioinnovation Centre (BIOZ) housing the Bioengineering Institute of TU Dresden and with space for commercial biotechnology under the same roof. We wanted BIOZ to be close to our institute in the Johannstadt district of Dresden. If BIOZ were built on the main campus of the TU Dresden, ten kilometres away, we would lose control of the project and be unable to recruit researchers of the quality we wanted. Besides, we would lose Francis Stewart, who was already onboard. He wanted to stay in Johannstadt at the centre of the Biopolis Dresden project.

The leadership of TU Dresden and most of its professors of biology and physics opposed our plan. They wanted to control the project. But we did not budge. If BIOZ failed, so would the MPI–CBG. We could not thrive without more activity around us. We wanted to establish collaborative projects in medicine and engineering in

BIOZ. It was a hard struggle, but with the support of the state government, we finally won.

One day, I was sitting in Wolfgang Vehse's limousine, telling him about our difficulties. Vehse was secretary of state at the Ministry of Economy and had supported our efforts to forge the new research environment. Without hesitation, Vehse called the chancellor of TU Dresden and started to discuss the issues with him. Finally, when he saw no other way to make headway with the building plans, he raised his voice. He, too, had learned the tactic that Harald Teir taught me in Helsinki. Hard talk. TU Dresden finally yielded. Our efforts had paid off. This victory was pivotal for the entire Biopolis Dresden project.

In addition to my efforts to implement our plans in Dresden, I had to continue my struggle on the European level to establish a funding programme for both junior and senior scientists. My credibility as ELSO president was at stake. The most important issue was the need to support young, independent researchers, who were the future of research in Europe. After being trained in research as a doctoral student and postdoc, a promising young researcher faced a career problem: there were few suitable job options. Finland and Sweden had already established programmes that funded positions for junior research group leaders. I was fortunate to be awarded such a position when I returned from New York. Elsewhere in Europe, however, postdocs usually returned to the lab where they had started their training, and there they continued to work for their professor. This meant that European researchers were often deprived of the chance to do fully independent research during the most creative period of their lives. Most fundamental discoveries are made by young investigators. In the USA, a functioning career ladder had been established, but in Europe, research was still dominated by authoritarian professors. There were many exceptions, of course, but they only proved the rule.

EMBL, EMBO, and ELSO led talks with the European Commission to establish a funding agency that would support fundamental research by individual excellent researchers, in addition to the multinational consortia funded by the established framework programmes. At one stage, the plan was to fund only senior, established researchers. That would have been a total disaster. To mobilise scientists to the cause of young investigators, ELSO started a campaign demanding that junior research positions be included in the new initiative.

Through The ELSO Gazette, Carol Featherstone organised a petition that collected some 5,000 signatures. We not only presented our petition in Brussels but also contacted science administrators in all parts of Europe to impress on them the importance of this reform. The molecular biologists led the way, and other scientific societies and actors joined in. This great concerted action led to the founding of the European Research Council (ERC) with a big budget for funding researchers at all levels, from those just starting an independent career to established experts in all branches of academic research, from the humanities to natural sciences.

This new funding agency inaugurated a new era in European research, and the ERC has become a phenomenal success story. The concept of Euroexcellence is now a reality. An ERC Starting Grant provides funding for young researchers to perform independent research for five years. With this funding, the recipients can choose where in the EU they want to start their career. It was fortunate that we had established ELSO just at the right time and that the EU commissioner for research, Philippe Busquin, paved the way for the ERC with his dedicated support.

International panels select the recipients of the ERC grants. For several years, I was the chairman of a panel and saw how well these expert groups work compared with national committees. Selection truly depends on scientific excellence and not on national considerations and networks.

I had a similar positive experience later that led to establishment of the European Innovation Council. During my years as a member of the board of the European Science Foundation (2000–2011), its then president, Reinder van Duinen, was fighting for change in the rules of European and national agencies for funding start-ups. He wanted these agencies to provide grants that would fund start-ups 100%, following the example of the Small Business Innovation Research program in the USA. With the goal of generating innovative high-tech companies, this highly competitive federal program encourages small business to explore their technological potential and provides the incentive to profit from commercialization. Some of the most innovative US companies, including the biotechnology companies Amgen, Biogen, and Genzyme, received early-stage financing from the Small Business Innovation Research program.

Reinder van Duinen's idea was making no progress, so I decided to get involved. The EU should copy successful programs. Following the model of its outstanding ERC – especially the Starting Grants for young investigators, which are a model for the world – I thought the EU should start an agency similar to the US's Small Business Innovation Research program for start-ups and early-phase companies. Panels of entrepreneurs and industry experts, as well as scientists, should evaluate the applications. Like the ERC, which has panels in all areas of basic research, the panels should cover all areas where new companies may be founded. Added value would accrue from bringing together experts from all over Europe to serve on the panels: they would learn from each other how to evaluate business plans.

I wrote a short proposal along these lines that I started to distribute to ministries and anyone I thought might contribute to achieving this idea. Most importantly, I had the opportunity to speak at a small workshop, organised by the Portuguese prime minister Pedro Passos Coelho, where I presented the idea. I also persuaded Leonor Parreira, the state secretary of the Portuguese research ministry in Lisbon, to present my proposal to the ministers of research of the Mediterranean EU member states, and she continued to support the idea. Amazingly, a new structure appeared in the European Commission's Horizon Europe funding programme: the European Innovation Council (EIC), which was very similar in structure to the one I had proposed in Lisbon. The EIC idea was put forward by the European Commissioner of Research, Science and Innovation, Carlos Moedas, who had been nominated for this post by Coelho and was previously secretary of state under Coelho in his Portuguese cabinet. After a pilot phase (2018–2021), a fully-fledged EIC was launched in 2021 with a budget of ten billion euros.

The EU's research policies are indeed functioning well. What a fantastic continent, despite all the negative press that the EU suffers. Unity in diversity is, by far, for me, the most awesome political battle cry.

***

Since I started my research career in the late 1950s, the international research community has grown tremendously. There are now some eight million active researchers in the world. In other words, the equivalent of about 90% of all the scientists who worked in the twentieth century are active right now! Many countries that previously had hardly any researchers are now represented in our global army.

China is a good example of the global growth of research. In 2006, I became involved in the problems faced by a country expanding its research culture. I was acting as the co-director of the Shanghai Institute of Advanced Studies of the Chinese Academy of Sciences. The MPG and the Chinese Academy of Sciences ran this institute together. It was intended as a think-tank for the academy. For me, the task was only a side job, but I was interested to see how the Chinese research landscape was developing and wanted to contribute if I could. I met the president of the Chinese Academy of Sciences and asked him what he wanted the Shanghai Institute to do. He answered that it could continue to operate as before and that we had full freedom to shape our activities. That was all. He gave no further instructions.

The most important activity at the institute was organising round-table conferences. It was located in a beautiful building that had formerly housed the French Consulate in Shanghai. Now it was the neighbour of the Science Academy's skyscraper for biological research. We would have no difficulty attracting participants. We started by organising conferences on nanobiotechnology and neurobiology. I insisted that we summarise our conclusions in short memoranda to be sent to the academy as food for thought to encourage new activities.

I convened young group leaders who had returned to China from abroad to discuss their experiences working in their homeland again. It was clear that they were worried about their future. Their biggest worry was that they did not understand the criteria by which research grants were distributed. The Research Ministry administered the largest budget, but their decision-making processes were not transparent. I decided to take the bull by the horns and organise a round-table discussion on the subject of 'Best Global Practices for Research Funding'. To give the discussion weight, I put together an illustrious panel including the MPG's president, Peter Gruss, the president of the USA's National Academy of Sciences, Bruce Alberts, and other influential scientists with personal experience in research funding. We couldn't proceed without the go-ahead from the president of the Chinese Academy of Sciences, however, and he remained silent.

Instead, there came a message from the academy requesting that the MPG increase its share of the budget for the institute. This was just before the Olympic Games in Beijing. I understood that, with my new initiative, I had crossed a red line. The academy was not interested in this important topic.

I decided to renounce my co-directorship. There was no point in proceeding if we did not have the Chinese Academy's support for important projects. I began to realise that the think-tank was mainly a cosmetic exercise. The political pressure from above was increasing, and since Xi Jinping took over, it has only worsened.

I wonder whether it is possible to build constructive and open research environments in societies like China's. In China, every institute is monitored by a representative of the Communist Party. Can a junior researcher contradict his boss without risk of retribution? If this is impossible, the control from above will inevitably quench creative and open discussion. That would not bode well for research in China. In their recent book 'The Power of Us', Jay Van Bavel and Dominic Packer stress how dissent can improve innovation, creativity, and group decision-making. "Dissent is effective because it changes how other people think. This means that dissenters do not actually have to be right to benefit the group – they just have to speak up to get others thinking." The complexity of research makes critical input crucial. We need critical feedback, even on our pet ideas.

In 'The Science of Liberty: Democracy, Reason and the Laws of Nature', Timothy Ferris writes that creative science cannot thrive in dictatorships. This has been the case thus far. Without freedom, research petrifies. We don't know whether Ferris will be right about China, however. Will China become an exception? Only the future development of Chinese research will give us the answer. Personally, I am inclined to agree with Ferris. Despite the huge investment in research in China, it's unclear whether the country will be able to make its research environments as creative as the best Western examples. Will China deliver discoveries and inventions in the same way as the

Western world has done until now? Will the country become a new motor of technological progress now when we desperately need new solutions for our planet?

Maybe the enormous resources given to science and technology will overcome the hurdles imposed by the Communist system. It is impressive how China has managed to build a functioning research system in record time. The publication record of Chinese scientists is constantly improving. I remain sceptical, however. In my opinion, the dictatorial policies of Xi Jinping do not bode well for the future. China does not meet the criteria of an open society that not only provides freedom for scientific discourse but also handles mistakes that will be made when introducing new technology. These problems are not foreseeable. They are what economists call unintended consequences. As researchers, we are doomed to make mistakes, but we can learn from them, correct them, and struggle.

***

India, also, has worked hard to improve its research base, but at a much slower pace than China. I once participated in a meeting with India's research minister, Kapil Sibal, at the invitation of Richard (Rick) Klausner, who acted as advisor to the Indian government. I had known Rick since he was at the NIH, and I had functioned as his external mentor. He had a stellar career and, at an early stage, became director of the National Cancer Institute with a budget of several billion dollars. Later, he led the Bill Gates Foundation before he became an investor in biotechnology. Rick had also invited David Baltimore, Eric Lander, and Gary Nabel to New Delhi. India had decided to accelerate development of its research and wanted advice on how to build elite institutions. David had established the Whitehead Institute at the Massachusetts Institute of Technology. Later, as president of Rockefeller University and the California Institute of Technology, he gained valuable experience leading successful institutions. Eric Lander founded the Broad Institute at the Massachusetts Institute of Technology, which was a leader in genome research. Gary Nabel was director of the newly founded Vaccine Research Center at the NIH. This group would now advise the Indian government.

We were all sitting at a table when Research Minister Sibal entered the room. To my surprise, he greeted each of us by our first names. This made a good impression on me. For two days, we discussed with representatives of Indian research institutions and laboratories. Above all, we counselled them that it is important to advance slowly. Led by Rick, we stressed to Sibal how unanimous we were on this point. Elite institutions cannot be conjured up instantaneously. In biomedical research, it is best to start with a single centre and proceed only after seeing how the first one takes shape. The most crucial issue, we argued, is to find an excellent director: someone who is not only a prominent and well-connected researcher but who also has a generous nature. Egoists have difficulty attracting talent and building environments that function well.

I almost fell off my chair when I realised that these heavyweights from the American research community thought exactly like me. Here, there was no talk about the Harnack Principle. In addition, we all agreed that young scientists are the basis for building excellence. Also, the choice of research fields is important and depends on the interests of the director. It was a fascinating meeting, which confirmed much that I had learnt during my years as research leader. I expanded our recommendations to include cooperation as a basic ingredient in building viable research environments. Unfortunately, our advice left no deep impression on the Indian research system, and the Indian government did not fulfil its plans to support Indian research as dramatically as it seemingly intended.

***

Another big change since I began doing science is that there has been a frightening increase in the number of articles published. New journals are constantly appearing in order to accommodate the flood of papers. But getting publications into the top journals has become a torture. This is the most negative change for me. It is frustrating waiting for referee reports, which we have learned to expect will be negative. The critiques can be malicious, requesting additional experiments that may take a year to perform but do not necessarily improve the article significantly. Even if the experiment is performed, the paper may still be rejected. This humiliation, which has become the new normal, is disheartening for young researchers.

Biology itself has become more complex: we know so much more than we did sixty years ago when I started. Now, we must come to grips with complexity. How can we produce results that are not meaningless because we have lost sight of the context? Making progress is an eternal struggle. Immunologist Peter Medawar said that science is the art of the soluble. By this, he meant that the art of research was making difficult problems soluble by devising means of getting at them. It is about finding answers to the *right* questions. How can we improve the process to find problems that are not only solvable but also are worth solving? Unfortunately, much of what is published is more or less worthless and immediately forgotten.

Nevertheless, it is exciting to be a researcher. The brief moments when you discover something new are unforgettable and spur you on. Opening doors to new vistas of scientific exploration is exhilarating. To add a piece to the puzzle that you are studying and suddenly see a new aspect of the whole picture is so awesome that it compensates for all your struggles. Insight does not always come as a sudden *Eureka!* moment, however; often the picture emerges slowly. This slow process is also fascinating. Every generation of researchers faces new challenges. There will never be an end to fundamental research problems.

One of the obligations of every successful researcher is to try to convey this fascination in the most captivating manner possible. Can we narrate our explorations so they catch the attention of the wider public? Can we tickle the curiosity that all children have but most adults have lost?

At the MPI–CBG, we invited internationally renowned researchers to give lectures that were open to the public. Bruce Alberts gave one such lecture describing the National Science Education Standards guidelines, which the USA's National Academy of Sciences had developed to reform science teaching. The new guidelines were undoubtedly excellent, maybe the best in the world. The academy had mobilised all its expertise to shape new curricula and plan practical approaches for teaching science. When Alberts retired from the presidency of the academy in 2005, he received the honorary title of 'Education President'. He had made a great contribution to improving science education in the USA.

But it was depressing to hear about the fate of Alberts' great reform. The project found scant support in the various states. It was a catastrophe. Only 17% of schools in the USA adopted the guidelines. It shows how little resonance science has in the American population at large, and this has had fatal consequences. The academy's guidelines encountered opposition from all directions. We only need to consider how large a proportion of the US population does not believe in evolution to realise what kind of obstacles the project had to overcome. The SARS-CoV-2 sceptics, who, to this day, do not believe the virus exists, are another example. If all Americans had received science education on the level offered by Alberts' plan, maybe the USA would be a better informed and pragmatic society. Naturally, the same principles also apply to Europe.

Bruce Alberts told us about an idea to stimulate curiosity about nature in kindergartens. I want to share one of his examples of how easy it can be. The children are asked to put on white socks and are sent out in the yard. The socks collect small particles and seeds from the ground. When the children come back inside, they pick off everything that has stuck to the socks and, using a microscope, they sort the particles into different heaps according to their shape. (A simple microscope costs less than ten US dollars.) To find out what the particles are, in the next step of their exploration, they plant them in flowerpots and see what happens. In this easy and tangible manner, the children learn about the workings of nature.

Alberts also emphasised how important it is that researchers talk about what they do. We have followed this credo in Dresden. We cannot hope for support from the government if we do not inform the public about what we are doing. Our public relations chief at the MPI–CBG, Florian Frisch, proposed the idea of a Science Café, where scientists discussed various contemporary issues in a city pub. The topics included, for example, genetically modified organisms in agriculture; stem cells; organ transplantation; animal experiments, and prolonging lifespans. These discussions demonstrated how worthwhile it is to engage the public: the pub was always full.

The MPI–CBG group leaders, postdocs, and doctoral students regularly gave lectures to senior citizens, schools, and students about the themes of our research in Dresden. Narrating what we are doing is not always easy, however. Making the stories intelligible and exciting requires practice. Also, we invited various groups to the institute for guided tours. Even kindergarten kids were welcome. Children are so incredibly curious. It was a treat for them to see our zebrafish and salamanders, and to use our microscopes.

With our support, Daniel Müller at BIOZ organised a yearly event at Dresden's Semper Opera, at which a prominent researcher gave an account of his field under the general title *'Zukunft heute'* ('Future Today'). A scientific lecture in this beautiful opera hall was a unique experience for both the speaker and the audience.

Every year Dresden organised a *'Lange Nacht der Wissenschaften'*, or 'Long Night of the Sciences'. All research institutions in Dresden opened their doors and arranged displays showing what went on inside the walls of the institute. We usually had some 1,200–1,400 visitors at the MPI–CBG. The enthusiasm of Dresden's citizens for research was inspirational and showed that our work was not in vain. In 2006, Dresden was elected the 'City of Science' in Germany.

The greatest achievement for us was that TU Dresden was successful in the Excellence Initiative of the German government, which allocated two billion euros to strengthen the German university system. In the first round of the competition, TU Dresden received funding for an 'excellence cluster' in regenerative medicine and a doctoral programme in biomedicine and bioengineering. In the next round, it was one of eleven universities appointed 'excellence universities'. Wieland and Ivan did an impressive job of helping the university to this overwhelming success. Together with TU Dresden's Rector Müller-Steinhagen, Wieland came up with the idea of a research alliance of TU Dresden with Dresden's non-university research and cultural institutions. A total of thirty-three partner institutions joined forces to develop the Dresden science hub into what was named by the new Rector of the university, Ursula Staudinger, DRESDEN-concept Science and Innovation Campus.

TU Dresden and Humboldt University in Berlin were the only East German universities rewarded in the tough, national Excellence Initiative competition, which was evaluated by international committees. None of the traditionally strong fields of engineering at TU Dresden qualified. It was gratifying that the new areas of research the MPI–CBG introduced into Dresden – regenerative medicine and the physics of life – were a driving force for this success.

The research landscape in Dresden had changed dramatically. But the process is far from finished. The unique cooperative spirit that pervades the research has surely contributed to its success. Together, we are strong. Let's hope that this spirit will live on in the future.

***

Research is such an important driving force for our society that we must protect the integrity of science. Above all else, we must avoid repeating mistakes that have decreased public confidence in our research enterprise in the past. The ethics of research have become increasingly important to me over the years. As researchers, we have a heavy responsibility to protect science against fraud and dishonesty. Every new generation of researchers must be taught to sustain the integrity of the research process. Despite many pledges, we are far from having established procedures that work.

Near the end of my career, I was involved in an episode that showed how difficult it is to sort out scientific misconduct. As chairman of the scientific advisory board of the Finnish Institute of Molecular Medicine, I and a group of experts was tasked with evaluating a lipidomics laboratory at the VTT Technical Research Centre of Finland. The centre was in the process of reorganizing and wanted to move its lipidomics group to the Finnish Institute of Molecular Medicine (FIMM). Our task was to assess whether the group met the institute's international standards.

While we were preparing for our visit, I received a phone call from a member of the lipidomics group. He told me that they had problems with their leader, who published without consulting his co-authors. Data were added to papers without consultation. Obviously, this was alarming. I asked him to work with the other group members to write an account of the problems and send the report to me.

During the scientific advisory board's visit to the centre, it was evident that the lipidomics group was divided: one half was loyal to the group leader, whereas the other half was dissatisfied. We could not recommend the move to FIMM. There were two problems: the research was not of the required quality, and the group leader did not follow good scientific practice. Our decision was unanimous.

In a single day, our expert group could not delve deeply into the complaints against the group leader, but we recommended that the VTT Technical Research Centre take the criticism seriously and investigate whether the complaints were founded or not. Until then, the management had swept the problem under the carpet and refused to respond to the claims of the whistle-blowers. We pointed out that it was the duty of the management to investigate whether the group leader had misused his power as a supervisor and whether this had led to manipulation of the data or not.

The centre ignored our request. They insisted that the problem was due to personal conflicts and not to misconduct. Finland had established a Delegation for Research Ethics, but this body would not interfere because it had no right to meddle in the internal affairs of a research institution. I persisted, however, and, at last, the centre appointed an expert group that analysed one of the articles of the lipidomics group but did not interview the whistle-blowers. The experts also did not check how the analyses had been carried out. It was a total farce. I complained again, but the Delegation for Research Ethics would not interfere. The problem was over and done with as far as the VTT Technical Research Centre was concerned. If research fraud had been proven, they might have had to repay research grants of the order of millions of euros to their funding agencies.

What made me fight for the whistle-blowers was that the centre did not follow good scientific practice. It did not respect young researchers and ignored their complaints. I must insist that I do not know whether fraud occurred in this case or not. It was never investigated. Without a doubt, the group leader was misusing his position of power, but the centre did not admit to that. The management did not live up to its ethical duty. I had to give up. The Finnish system for protecting research integrity was simply not functioning. Finland could learn from Germany, which has a much better way of safeguarding good research practices. An important feature of the German system is the ombudsmann for scientific integrity. The ombudsmann is a gremium appointed by the German Research Foundation, that assists all scientists and researchers in Germany when it comes to questions and conflicts related to good scientific practice and scientific integrity.

The behaviour of the VTT Technical Research Centre is not exceptional. Whistle-blowers find it hard to make their voices heard, even though the research community usually affirms that it takes research ethics seriously and thoroughly investigates all complaints. Occasionally, dishonesty in research reaches the ears of the wider public. A great scandal at Karolinska Institutet in Stockholm a few years ago, for example, showed how sluggish institutions can be in their reactions. The Swiss-born Italian thoracic surgeon Paolo Macchiarini had been recruited as a visiting researcher and senior surgeon. He was an expert in stem cell transplantation and transplanted stem cell-coated artificial tracheas into patients at Karolinska University Hospital. All three patients died. Macchiarini turned out to be a charlatan who had faked his previous achievements. The leadership of Karolinska Institutet had succumbed to his charm. It took several years before the affair was resolved. All those involved were dismissed, and the reputation of Karolinska Institutet was severely damaged. This sad episode shows how even top-level institutions sometimes let themselves be fooled.

Because research plays such a great role in modern society, it is important that we take fake science and research ethics seriously. The integrity of research is one of the pillars of western democracies. We cannot accept falsified or unethical research. The reputation of all that we stand for as scientists is at stake. Society depends crucially on an intellectually free and functional research system. The insights delivered by science must be subjected to permanent checks. This process never stops. The validity of scientific data is continuously scrutinised and must be defended with factual arguments. If new data overturn earlier conclusions, those conclusions must be revised. As Holden Thorp, editor-in-chief of the Science family of journals wrote, "Science is work in progress."

Innovations that build on false data may be disastrous if they are not corrected. To hammer home this point, I want to cite two warning examples.

Trofim Lysenko was, arguably, the researcher who caused the most damage through fraudulent science. During the 1930s, he started a campaign against 'Western' genetics in the Soviet Union, where genetics, with Nikolai Vavilov as its foremost representative, was of international repute. Like Lamarck in the nineteenth century, Lysenko proposed that acquired characteristics could be inherited. He had no real evidence for this thesis, which was more or less made up.

The theory was disastrous because Joseph Stalin, the Soviet leader, became fascinated by Lysenkoism. This new doctrine held the promise that a successful communist education would be passed on to the next generation. With Stalin's support, Lysenko managed to dispose of all the leading geneticists of the day and began to reform agriculture according to his own ideas. The result was a catastrophe: harvests decreased, and famine struck. Lysenko remained in power even after Stalin's death in 1953 and was pushed aside only in 1962.

Chairman Mao Zedong of China, too, let himself be fooled by Lysenko's heresies. Mao had read about his revolutionary agricultural methods and their spectacular results. He was so fascinated by Lysenko's ideas for socialist agriculture that, during the 'Great Leap Forward' in the late 1950s, he ordered the introduction of socialist methods according to Lysenko's system.

Fortunately, the directives did not reach all of China, but a great part of the country did follow them. When harvest time came, Mao was delighted by the overwhelming reports about increasing harvests that were sent to him. The reports were falsified, however. Lysenko's methods led to a great famine. Almost thirty million Chinese died. The famine was not due to Lysenkoism alone, but the consequences were disastrous for China.

Nazi Germany is another warning of how badly wrong things may go when pseudoscience takes over. Social Darwinism was established as an ideology at the end of the nineteenth century. Herbert Spencer coined the expression 'survival of the fittest' to encompass the concept that only the strongest survive in the struggle for life. This basic view affected the British naturalist and geneticist Francis Galton, who developed the idea of eugenics – the improvement of the genetic characteristics of humans through selective breeding. The underlying thought was that humans have liberated themselves from the struggle for life and natural selection and must, therefore, do the selection themselves. The weak must be prevented from reproducing.

This was pseudoscience of the worst kind. Spencer launched his concept without the support of Darwin, who did not agree with his interpretation. Darwin did not believe that the strongest always survive. By contrast, he saw group cooperation as a positive selection factor. According to Darwin, the success of *H. sapiens* in the struggle for life was due to intelligence.

This pseudoscience was wholeheartedly adopted by Hitler and his regime. Eugenics meant racial hygiene. The Aryan race was superior to all others, and therefore the Nazis had to make sure that the Aryan blood was kept pure and strong. Other countries force-sterilised their so-called inferior population groups, whereas the Nazis implemented their programme of racial hygiene by killing those they wanted to eliminate.

In his book 'The Nazi Doctors, Medical Killing and the Psychology of Genocide', Robert Jay Lifton describes how the German medical corps were so brainwashed that many took it as the duty of a physician to participate in this killing. Thus, the Nazi regime built a moral foundation for its eugenics policies. Social Darwinism has not died out. Even today, its interpretation of nature is poison for our society.

***

Benno Müller-Hill, a leading German molecular biologist, was one of my research acquaintances. He trained in James Watson's lab at Harvard and later became a professor in Cologne, Germany. In 1983, he took a leave of absence to write 'Murderous Science', a book about the German eugenics catastrophe. To document how it had happened, he interviewed scientists who had taken part, as well as their closest family members. An illuminating example is Eugen Fischer. He was the leading geneticist in Germany during Hitler's time and an active proponent of racial hygiene. Yet, nobody around him had a bad word to say about him. Everybody kept silent. Benno even found a letter that Fischer had written to Hitler, in which he profusely thanked him for making it possible for German geneticists to apply their principles of racial hygiene with such efficiency.

Initially, the German scientific establishment was not ready for Benno's book. He was ostracised and had difficulties continuing his ground-breaking research. This shows how painful it was, even fifty years later, for the German scientific elite to acknowledge the heavy involvement of science in the Nazi calamities. Later, Benno's efforts compelled German research organisations to scrutinise their activities during those times. This late settling of accounts was important for improving the research climate. It became part of Germany's road to redemption. No country has done more to atone for its disastrous past. We invited Benno to the MPI–CBG to give a lecture about his experiences. He also participated in a Science Café discussion. He made it clear how, ultimately, it is up to each of us to live up to the ethical expectations that ought to guide our research activities.

Today, we know that racism has no scientific foundation. Modern genetics gives no grounds for dividing ethnic groups into races. The genetic constitution of people of European, Asian, African American, and South American descent is 99.9% identical. Genetic differences within each population may even be greater than those between populations, depending on their geographic location. Ultimately, there is so much similarity between the races and so much variation within them that two people of European descent may be more genetically similar to an Asian person than they are to each other. Nevertheless, this 0.1% difference has led the world into calamities.

The plain fact is that the privileged position of the 'Aryan race' in Hitler's Germany had no basis in science. From a historical perspective, this may be surprising, but it is true. Hitler's Aryan race concept has been dismissed to the garbage dump of history. Differences in skull type and skin colour led to the cultivation of a concept of race that finished in disaster. The concept is obsolete and doomed to disappear from public consciousness. With this insight, we may build a more equal world. How can we, as scientists, persuade everyone of this truth? The fallacy still lives on.

After the calamitous first half of the twentieth century, West Germany established a deep-rooted democracy. In Heidelberg, we experienced how well the school system had come to grips with the difficult history of Germany.

The situation in East Germany was different. There, they thought that most of the fascists were in West Germany and, therefore, East German society did not have to engage with the past in the same way as the West. Anti-fascism was a pillar of the constitution of the GDR. The illusion that the Nazis had gone west was false, however. Had this been the case, the GDR would have been dangerously depopulated. Dresden, for example, was a Nazi stronghold. This became evident once again when the anti-Islam, far-right Pegida movement started in Dresden.

The East Germans have had more than their share of misfortune compared with other people in Europe. Like the rest, they suffered the horrors of the First World War. Then came Hitler with his horrible regime. Nazi Germany started the Second World War, plunging the world into deadly chaos with consequences unprecedented in history. After the peace treaty, the people of West Germany were able to build a new society based on freedom and prosperity. The East Germans, by contrast, were forced to turn to communism.

This painful transformation almost drove the GDR to bankruptcy forty years later when, in 1989, the wall fell. West and East were reunited. Now, the poor East Germans had to turn their coats yet again and adopt the unwritten laws of capitalism. Their lives were once more turned on their heads. For many, the transformation was much too brutal. Living through two dictatorships with different ideologies left deep marks. After the initial rejoicing over the reunification, the euphoria turned into a hangover.

Many East Germans were especially annoyed by the arrogant, *besserwisserei* (know-it-all) attitude of many West Germans. This resentment found expression in demonstrations by Pegida every Monday in Dresden. At first, there were a few thousand demonstrators. Then 5,000, 10,000, 15,000, 20,000 and 25,000 people crowded the city centre. The leader of the movement, Lutz Bachmann, was a criminal with sixteen burglaries, drug trafficking, and acts of violence on his record. He was an arch-right-wing activist with fascist sympathies. This background should have warned most people against assembling around him and other equally extreme agitators. But as the movement grew stronger, more people joined. Pegida gave resentful Easterners an opportunity to vent their despair. They felt that the rug had been pulled from under their feet. They had to find someone to blame for the misery they suffered. The most convenient scapegoat was, quite naturally, the West Germans. The more time passed since the GDR period, the more blurred their memories became of what socialism had actually been like. What people now remembered was that there had been jobs, food, and housing at reasonable prices. Above all, people had felt respected during the GDR period, and this was what they missed. This loss was not easy to compensate for.

Although only a minority of Dresden's population supported Pegida, the situation was unpleasant and worrying. Together with a group of friends, Carola and I participated in founding the 'Dresden Place to Be!' society. With this organization, we wanted to support and help foreigners who came to Dresden. Most were researchers and their families. Together with Gerhard Ehninger, a socially active professor of internal medicine, we came up with the idea that Dresden Place to Be! should organise a counterdemonstration. Gerhard mobilised his contacts and devoted all his energy to the organisation of this action. The event took place at Neumarkt in front of the *Frauenkirche*. He engaged one of Germany's best-known singers and musicians, Herbert Grönemeyer, and his band. This made it possible to attract many other musicians as well. Twenty-five thousand Dresdeners turned up! The atmosphere was fabulous. The musicians sang for Europe, internationalism, and solidarity, and the audience was thrilled. The event was a colourful contribution to the struggle for a more open world. When Grönemeier shouted into the microphone that everyone is for everybody in Germany, the crowd went wild. What a memorable occasion!

***

I was entering the home stretch of my research at the MPI–CBG. Progress had been slow with our research on the dynamic nanometre rafts, our main theme. My brother Tom once asked me if I experienced periods of depression when my research did not advance and I was going around in circles despondently. My honest reply was that I did not. Creative artists who work alone are more prone to attacks of depression. I worked in a team, which meant that there was usually someone in the team who saw some glimmer of light at the end of his or her tunnel. When I was completely stuck, I engaged in some project that had nothing to do with my research – ELSO or the Shanghai Institute, for example – to get started again. The risk of this strategy was that I would lose my research focus. This balancing act dominated my professional life.

Fortunately, light was now appearing at the end of my tunnel. I had managed to assemble a strong team around me, attracted by our research and the desire to understand how cell membranes compartmentalise to form dynamic raft domains. We were now moving out of membrane trafficking and focusing on the membrane sub-compartmentalisation concept that I had developed.

I have always been aware of how important methodology is for solving difficult problems, and my team members expanded our experimental repertoire with new vigour. We tried mass spectrometry for lipid analysis and imaging methods using various probes to identify rafts. We took membranes apart and reconstituted them again. We studied how lipids interact with membrane proteins. We studied bacteria to identify predecessors of the key raft lipid, cholesterol. New, high-resolution imaging methods allowed us glimpses of the elusive resting state of the rafts before they become activated. The collaborative atmosphere in the lab was great. As I was going to retire soon, my young collaborators knew they would be able to take their research with them when they started their own groups. They were building their own future research themes. Some of them have already realised their dream and are now leading researchers in their field. It was a privilege for me to enjoy such an exhilarating, creative atmosphere during the home stretch of my long career.

How can a thin two-dimensional fluid sheet be partitioned into minute 'reaction chambers' that soon disappear again? The reaction chambers form only when they are needed in response to a specific activation signal. During their transient existence, rafts function in a variety of processes in the cell. Their most important task is to create isolated spaces in the membrane where enzymes and other proteins can function undisturbed and in parallel to one another. But that's not all. The lipids in these nanodomains interact with proteins, modulating their activity. Cell membranes have remarkable material properties.

In the days when the idea of rafts was still under attack, I applied for a research grant from the ERC that was turned down. This rejection felt excruciating. Why was I not given this opportunity to complete my research career? I had given so much of my time to Europe. Might I not expect a bonus for that? Stupid of me, of course; our granting system does not function like that. Today, I would like to send a report to the reviewers of that grant application showing what we have been able to achieve. In the comments on my rejected research plan, I read that my project was too ambitious to be feasible. They simply did not believe that rafts existed. I am immensely happy that I managed to fund my research, even without support from the ERC. Most satisfyingly, nearly all the goals I set out in my application were achieved.

The breakthrough came when we and others were able to show that the dynamic organisation of rafts we proposed is based on a physical principle: a liquid–liquid phase separation. An everyday example of the physical principle of phase separation can be seen when we mix a salad dressing of vinegar and olive oil and let it stand: the oil and vinegar separate. Like oil and vinegar separating, the formation of rafts is driven by the separation of two liquids.

This phenomenon was already known for simple mixtures of lipids but was thought not to apply to the mixture of hundreds of different lipids and proteins in biological membranes. But it did. Before phase separation, the membrane lipids and proteins float around in a single phase. After local phase separation, the raft lipids and proteins form separate membrane domains that are liquid, where the lipids are more ordered and condensed or, in other words, more tightly packed. This difference in order explains why the rafts separate from the rest of the cell membrane, where the lipids are in a more disordered liquid phase. The sub-compartmentalisation process is dynamic: the rafts form and dissolve again. Rafts are not only isolated spatially but are also endowed with distinct material properties that arise from the molecular properties of the lipids that give rise to phase separation. Importantly, we demonstrated that these sub-compartments form when protein ligands trigger multivalent associations of lipids. Of course, we were not alone in this field. Other researchers added their pieces to the puzzle. It was incredibly satisfying that, after all these years, we were able to convince ourselves that rafts were dynamically robust in our cell membranes. Hooray for them! What a relief.

Another discovery about rafts that delighted but also intrigued me was their collective chemistry. The lipids in rafts have a weak affinity for one another: they associate to form cooperative collectives. These collective properties make rafts functional and give them capabilities beyond their individual constituents. I was amazed and amused that the principle that had pervaded my own life had also invaded my research. Togetherness adds value, and *H. sapiens* has evolved with this ability to cooperate in a collective as a driving principle for the continued existence of our species on this planet.

This insight dawned on me while I was writing this book and made me euphoric. Collective principles turn up everywhere in molecular biology. The chemistry of life involves affinities between macromolecules and small chemicals, allowing them to associate spontaneously to form collective entities with the biological properties required for life. The principle of self-assembly was a driving force during evolution, underlying molecular modules and units that were used repeatedly in all cells and organisms. Rafts, which are self-assembling structures in cell membranes, provide just one example of this general principle.

Looking back, I am painfully aware of what a drain on me it was to muster the strength that I needed to navigate my rafts through the counter-currents of criticism. My faith in myself was generated by the conviction that the basics of the raft concept were correct. Without this self-assurance, it would have been difficult to survive scientifically and to convince my co-workers to dig deeper. Only by digging deeper could we keep the rafts idea afloat. Naturally, I trembled every time a paper came that tried to refute the idea. But then, when I saw the data that was supposed to repudiate our work, I calmed down. The attempts to disprove our idea still go on, but also, nowadays, the supportive papers get through. What a relief after all the hassle. Every pioneer trying to open up new space within their territory of activity must go through these periods of agony.

Today, when I review what we have achieved, I realise that we might have pushed ourselves even further. I knew that we should have clarified the specific structural features that characterise the proteins in rafts. What is the basis of their specificity? What are the characteristic features of membrane proteins in rafts? What distinguishes them from membrane proteins that are excluded? Unfortunately, I did not give these questions the priority they deserved. As long as we had not demonstrated that such specificity existed, the critics would go on vilifying the poor rafts. We really should have filled this gaping hole in the evidence.

In 2017, Ilya Leventhal, one of my American postdocs, solved the problem of how a transmembrane protein becomes raftophilic. The solution was so simple that we should have thought of it ten years ago. The transmembrane domain of the protein is very slim, so it can enter raft nanodomains, whereas proteins in the more disordered segments of the bilayer membrane have much thicker transmembrane domains. In terms of their protein chemistry, the amino acids that make up the transmembrane domain of raft proteins have only small side chains, minimising the surface area of the domain, whereas in proteins that do not enter rafts, the amino acid side chains are much bulkier.

***

Phase separation has become an entirely new principle to explain cell organisation. Together with his postdoc Cliff Brangwynne, Tony discovered at the MPI–CPG that the cytoplasm contains condensed fluid droplets, which also form by phase separation. Whereas rafts form by phase separation in two dimensions in a cell membrane, these condensed fluid droplets in the cytoplasm and in the cell nucleus arise by phase separation in three dimensions, like the morning dew on the grass in late summer. Just like rafts, they form reaction chambers for various cell processes. However, as was the case with rafts for a long time, one key issue is not yet clear: what biochemical events specify the formation of different condensates? Life processes always involve both physics and chemistry. Combining the two remains a tough nut to crack.

These new discoveries give us completely new perspectives on cell organisation. The more we learn about the biology of the cell, the more clearly we understand how physical principles provide the chemistry of life with fascinating properties far beyond what one might expect from those of the individual chemical components. These physical and chemical interactions give the collective its attributes. Karl Marx must be smiling in his heaven. We call this ability of the whole to surprise us with new functions, the 'emergent properties' of biological systems. Above all, the dynamics and the specificities of cell organisation are striking. If we could swim around in the cell's interior and observe the structures that appear and disappear, we would see an unrivalled spectacle. The cell is full of reaction chambers, where chemical processes are separated in nanometre spaces to boost their efficiency. I'd love to know what other new principles of cell organisation will emerge in the future.

All this sounds interesting, a lay reader may say, but what is this research about phase separation in membranes and in the cell interior good for? Our insights into how cells are organised led us to collaborate with Andrej Shevchenko to develop a quantitative mass spectrometry platform for membrane lipid analysis at the MPI–CBG. We hope that this platform will provide the basis for diagnostic tools that measure health and disease. These ambitions may come to nothing, of course. Nevertheless, it is my firm contention that the global knowledge base of biomedical research constitutes a treasure trove for future applications.

In collaboration with the Massachusetts Institute of Technology, Tony founded a company called Dewpoint, that attracted an astounding starting capital of sixty million dollars. The company's activities are already underway in Dresden and in Cambridge, USA. The pharmaceutical giant Bayer has invested another 100 million dollars, and other companies have followed. The goal is to develop new drugs that interfere with processes driven by phase separation in the interior of the cell. Maybe Dewpoint will also include phase separation in the cell membrane in their remit?

The start-up company Jado, which Marino, Temo, and I founded to develop drugs to target lipid rafts, ended in failure. It was founded as part of our campaign to demonstrate to the Saxony state government that we care about biotechnology. But the timing was bad; it was too early to commercialise rafts. Jado went bankrupt in the aftermath of the 2008 financial crisis.

Basic science is almost always way ahead of successful applications. Sometimes, it is difficult to identify where such useful outcomes hide. This is the fascination of innovation. Society cannot decide beforehand what areas are important and which ones are not. Innovative breakthroughs may pop up anywhere. I am convinced that the research and development programme that Jado launched will be revived when pharmaceutical companies realise that cell membranes, especially the plasma membrane, are great targets because they are more accessible than the cell interior. Since the many different rafts operating in the membranes arise by specific activation, each activated raft is potentially a target, opening up a totally new chemical space for drugs.

***

The time was now ripe for me to decide what I should do after I closed the doors of my lab. I had counted on devoting my time to Jado when I left research, but that had foundered. So, what should I do now? Carola suggested that I should finally discover slowness. Would I be able to adopt a slow lifestyle? This would be more than difficult. When I'm idle, I feel a physical restlessness that seeks an outlet. It is not unpleasant; it is what drives me to think ahead. Over and over again, I juggle different ideas. When I discover I am left with a losing card, often I already have a new one in my back pocket. It was definitely too early for me to try slowness.

Günter Blobel invited me to an exclusive farewell symposium in Gstaad, Switzerland, hosted by the food and drink company Nestlé. Günter was a member of the Board of Directors of Nestlé and was about to leave his position. My lecture was about lipids and membranes. After my seminar, I realised that Nestlé had thrown out lipids from their research programme. I asked Werner Bauer, who oversaw the company's research and development and who attended the seminar, how such a decision was possible in a food-producing corporation. He admitted that it was unwise.

His reply gave me the idea of trying to persuade Nestlé to establish a lipid research unit in Dresden. At first, Bauer was ready to support my idea. Gradually, however, I realised that the only option was for me to find another start-up. I had earned money on stock options from Biogen as compensation for my work, which meant I had the funds to invest in new biotechnology. Why not?

Carola thought it was a good idea. It felt like a new beginning at the age of seventy-five. I had read that the Swedish premier Fredrik Reinfelt advocated that fit septuagenarians, with their rich experience, should start enterprises. Maybe I was not totally mad.

We named our new company Lipotype. Lipidomics is the collective name for everything that has to do with lipids in biological systems. Lipotype would market the mass spectrometry technology we had developed together with Andrei Shevchenko at the MPI–CBG. We wanted to offer a service to researchers in academia and industry worldwide for analysing lipids in cells, tissues, and body fluids. Our method had the potential to be commercially viable in this market, and we wanted to develop the technology for use in healthcare, too.

The research community's knowledge about lipids virtually disappeared during the DNA revolution, and other fields like metabolism and physiology were also marginalised. When we started, Lipotype had no serious competitors. Our goal was to prove the need for lipidomics. We must convince researchers in academia and industry that lipids are important. Let us see whether we succeed.

At the same time as Lipotype was taking form, the German Research Ministry announced a big new project for eastern Germany. It was called Twenty20 (*Zwanzig20*, in German), and the aim was to identify projects that could, in collaboration with industrial partners, establish a new technology with commercial potential. Each selected project would be granted twenty-two million euros from the government, and industry partners would contribute another twenty-two million euros.

This time, I had to assemble the consortium and put together the proposal on my own. The last research project in my lab had been supported by funding from Berlin that was aimed at starting a new company. Those who were active in this project, above all Oliver Uecke, supported my Twenty20 initiative. In addition, my daughter Katja, who had moved to Rome with her family for two years, agreed to write our project application in German. The project focused on healthy living, *GesundLeben*. Lipotype would play a major role in the project.

This was an entirely new field for me. The huge research appropriation of *GesundLeben* allowed me to fish for collaborators who would never have taken a smaller bait. It was an incredibly hectic time. We assembled a large consortium from academia and industry, which was ready to build new strategies for better health. We had chosen to concentrate on obesity, which is spreading like a pandemic all over the world. Today, obesity is regarded as a life-long disease. Forty percent of the adult population of the USA is obese, and there are no strategies to stop it. This disease can be deadly, leading to type 2 diabetes, myocardial infarction, dementia, cancer, and fatty liver. In the USA, the obesity pandemic costs 350 billion dollars annually in healthcare and a staggering one trillion in loss of productivity. Health insurance companies all over the world are unwilling to fund treatments until obese people develop diseases – much too late. Prevention is the poor relation of medicine.

It is totally incomprehensible to me that this worldwide obesity pandemic has not received more attention. Obesity is all over the media, but the problem only spreads. It can only be solved by prevention, for which there is little funding. The major cause of this health problem is the western lifestyle. Our metabolic control mechanisms can handle only so much fast food, snacks, and soft drinks before they reach a tipping point. Inflammatory reactions begin that can cause irreversible damage. The pandemic of metabolic disease is far more severe than that of COVID-19.

Our megaproject *GesundLeben* proposed to use the 'omics' revolution in biology – genomics, metabolomics, microbiomics, and lipidomics – to define risk factors for use in obesity prevention programmes in Germany. Exercise and diet were also part of our prevention plans. We wanted to help overweight and obese people change their lifestyles.

The proposal made it to the final round of selection by Twenty20, but it was not selected. In retrospect, this was fortunate for me. Our consortium of ninety partners was far too large to be manageable. It would have been a gigantic job to define realistic subprojects. Carola was relieved by the outcome. There had to be some limit to the undertakings that my restlessness pushed me to embark on!

Still, our efforts were not in vain. In developing the *GesundLeben* proposal, I had learned a lot of useful information about what Lipotype might do. Obesity would become a research and development goal for the company. Fats and lipids in the diet play an important role in people becoming overweight. Today, Lipotype has over thirty employees. My previous co-workers – Oliver Uecke, Christian Klose, Julio Sampaio, Michal Surma, and Mathias Gerl – formed the core of the team that started the company. Our service for researchers in academia and industry around the world has been so profitable that we have resources to do our own development work.

Currently, the company is developing a blood test that can be done at home. You prick your finger, apply a drop of blood to a filter paper, and post it to Lipotype for analysis. Our research indicates that the blood lipidome provides a good measure of the state of your metabolic health. With this personified health measure, our customers will have the option to decide for themselves how to change their lifestyle to avoid becoming victims of the obesity pandemic. Potentially, we are filling a glaring gap in clinical diagnostics: no diagnostic test available today measures the state of your metabolism. Just think about it: we are in the midst of a pandemic of unhealthy weight, yet we have no warning test for those whose metabolism has gone awry. That would be the same as having had no SARS-CoV-2 test for the COVID-19 pandemic.

Perhaps Lipotype will live up to its slogan 'Lipidomics for a better life' and show that lipidomics can become a tool in the fight against this gruesome pandemic of metabolic disease, which is receiving too little attention. The high-tech giants develop smartwatches and other gadgets that give us information about our physical health. But for more precise information on our health, we also need chemical data. These Lipotype can deliver. My engagement with Biogen brought me insight into how a successful biotech company works. The experience from our start-up Jado has, I hope, taught me how not to proceed. Maybe Lipotype will become a success? Only the future will tell.

Epilogue

The COVID-19 pandemic showed how extremely vulnerable our world is and how much we depend on one another. The effects of the SARS-CoV-2 virus were felt in every corner of our planet. The pandemic clearly demonstrated the importance of science and research for our global well-being. The research community identified the cause of the pandemic in no time, and then rapidly initiated a global effort to develop vaccines against the virus. It was a triumph for research.

Who would have thought it was possible to produce new effective vaccines so rapidly? The investment was enormous. More than sixty vaccines have been or are being tested in the clinic and some have been approved for use, permitting vaccination programs all over the globe. The success of the new RNA-based vaccines is fabulous. Their efficacy is impressive and shows the huge potential of our global research community. But we scientists have also learned that it is not sufficient simply to deliver a technical solution to a problem. The vaccines had to be accepted by the population. Psycho-sociologists and other experts should have joined with the virologists to devise a common action plan. In fact, most problems are easier to solve when we approach them together with all the available expertise.

This applies also to a related problem that surfaced during the COVID-19 pandemic – the great number of people who not only openly declared themselves anti-vax but also anti-science, anti-Western classical medicine, and/or pro-QAnon or other conspiracy theories. This was such a surprise to me. How could anyone accept that the COVID-19 pandemic was not caused by a virus? This mishmash of superstition and fake science is utterly disturbing and requires serious thought by our science community. The upsurge of medieval thought in modern times cannot be neglected because the hodgepodge of ideas in their message will be propagated by all the channels that distribute fake news, using the new capabilities provided by artificial intelligence. This is a scary perspective for the twenty-first century. We must defend the foundations of our scientific culture in an age of misinformation.

It is a pity no one took Prince Kropotkin seriously. The twentieth century might have looked different if his altruistic message had had more political influence. Instead, it was dominated by the infernal message of selfishness, with *H. economicus* as its emissary. The rich have become richer, and the poor have become poorer. Capitalism brought with it the disturbing insight that it was possible to keep the population mute despite all injustice: industry only has to drench us in a torrent of consumer goods to blunt all resistance. The hysteria of consumerism has spread like wildfire over the world, engendering ruthless exploitation with disastrous consequences. It exploits both human labour and natural resources. Capitalism's law of 'cheap nature', as coined by Jason W. Moore, has taken us to the edge of a precipice. Earth is approaching a tipping point that we cannot transgress without fatal consequences.

We have known about climate change since the 1980s. The nearly unanimous voice of climatologists warned us of the consequences, but that was not enough to wake us up. The scientists did not understand that they had to expand their efforts to engage a broad enough collective of expertise worldwide to pull us out of our indifference and lethargy. As the world was finally awakening, the most important industrial nation had a president who defiantly counteracted all attempts to guide his country and the world towards a sustainable future. President Trump is not known as a reader, but he names Ayn Rand's novel 'The Fountainhead' as one of his favourites. The book's title is a reference to Rand's deplorable statement that "man's ego is the fountainhead of human progress." Rand's destructive influence extends far into the twenty-first century. Trump's view of science was equally destructive. Georg Henrik von Wright in his book 'Understanding the Future', wrote that the sum of superstition in the world is roughly constant. I could hardly believe my eyes when I read that sentence years ago. He added that in a scientific era like ours, superstition is dominated by scientific superstitions. Von Wright was right again. Every day of his presidency, Trump and his henchmen delivered stubborn and mendacious scientific claims that harmed us all. It was absurd beyond comprehension. Unfortunately, the influence of Trump's Republican party is still taking its toll on efforts to stall climate change in the USA and, thus, in the world. We can only hope that the Democrats will prevail again in 2024.

Worldwide many people, most of them young, are taking an active approach to climate change. In 2019, Greta Thunberg performed a miracle. Her movement forced politicians to include climate problems on their agenda. The EU countries agreed to lower their emissions of greenhouse gases by at least 55% before 2030. China promised to decrease its carbon dioxide emissions radically and to be climate neutral by 2060. This news awakens hope. Might we listen to reason after all?

On my most positive mornings, I wake up and imagine that I see the outline of a new world where selflessness, compassion and altruism are marching forward. A chorus of researchers and writers encourage us to change our course. For example, Matthieu Ricard's book 'Altruism: The Science and Psychology of Kindness'. Stefan Klein's 'Survival of the Nicest' is like a late sequel to Kropotkin's 'Mutual Aid', discussing how altruism made us human and why empathy pays off. Edward Wilson's 'Social Conquest of the Earth' is another important book. Wilson is a renowned investigator of ants and the evolution of sociality in animals. His studies have shown that there are only a few eusocial animals on our planet: bees, ants, mole rats and *H. sapiens*. Eusocial species live in groups and have a unique division of labour in their societies. The insatiable success of *H. sapiens*, altruistic by nature, has depended on evolving flexible group cooperation. Another example is 'Survival of the Friendliest: Understanding Our

Origins and Our Common Humanity', by Brian Hare and Vanessa Woods. Also, 'Blueprint: The Evolutionary Origins of a Good Society', by Nicholas Christakis, filled me with hope and warmed my heart.

Penguins show altruism, too. When the temperature drops to -40°C, they aggregate in tight huddles. Slowly, individuals move from the cold outer edge of the huddle towards the warm middle, and this dance continues for as long as the huddle persists. *H. sapiens* knows more than the penguins, of course. We are 'SuperCooperators' according to Martin Nowak, who writes about altruism, evolution and why we need each other in the fight for a better world. I agree completely with the essayist Tony Judt when he wrote in 'Ill Fares the Land' that the materialistic and selfish quality of contemporary life is not inherent to the human condition.

Another sign of change, reflecting a general trend, is taking place in the arts world. In 2021, the annual Turner Prize, one of the most prestigious awards for the visual arts, went to the Belfast-based Array Collective. In fact, all five finalists for that year's Prize were collectives! Might research foundations follow suit and accept that research today is performed by collectives of scientists?

Are we indeed starting to wake up from a period of shame for the human race? The twentieth century employed many of the technologies created by science for the purposes of murder and destruction. Not even high culture immunised the population against Hitler's madness. On the contrary, many artists marched along singing. Hitler managed to galvanise the German people to cooperate, but towards devilish goals. Maybe we can change direction and galvanise the world towards altruistic cooperation? To survive as a civilisation, we have no choice: we must build a sustainable and just world.

Finland is reforming its school curriculum to introduce cooperation and synergy. In the 1970s, reforms were introduced to make schools equal for everyone. The weaker students received as much attention from their teachers as the stronger ones. I remember so well the storm of criticism that the plans for these reforms unleashed. Common opinion was that the quality would drop with disastrous consequences. But the result was the opposite. Finland became a shining example for the world. Finnish industry benefitted, too. Without this school reform, the Nokia miracle would not have been possible.

Now, pupils and teachers in Finnish schools collaborate by working together in groups, and this collaboration includes the parents as well. This reform may be revolutionary. I hope that it is pursued with the same energy that characterised the previous reform. The Finnish female teaching staff played a decisive role in implementing the reform. An altruistic society is not possible without a strong female imprint.

We stand at a crossroads. We are forced to choose our way. The choice should be clear. We cannot go on as before. But even if we choose to move our economy and our society in a more equal, energy-efficient and sustainable direction, the change of direction, inevitably, will be slow.

It is not sufficient to introduce technological innovations; we must also change radically our lifestyle. I play with the thought that we need transitional stages on our way to a society that makes survival possible. Maybe we could create a sector of society that investigates and experiments with how we can live more sustainably? We must mobilise all imaginable innovations as well as social support to make the experiments work. We must add 'with sustainability' to the EU's motto 'United in diversity'.

At eighty-five years old, my vision for the future remains hopeful, despite all setbacks. *H. sapiens* will throw off the individualistic yoke that has brought us to this impasse. Even Putin's disastrous war in Ukraine cannot change my mind. There will always be evil in human society. But I believe the youth of this world will find their way back to our altruistic roots and learn to cooperate to build a better and more equal world. They simply must.

We should not abandon the rationale of the enlightenment. It has, after all, brought prosperity and well-being to large sectors of the global population. But we must learn to combine regional diversity with a common goal: sustainability.

I remain an incurable optimist. I wish I were twenty again to be part of the incredible, global experiment that I hope will materialise.

Scientists Working in the Simons Lab

1968 – 2013

HELSINKI | 1968 – 1975

Carl Gahmberg • Ari Helenius • Henrik Garoff • Rosette Becker • Gerd Utermann

HEIDELBERG | 1975 – 2001

Henrik Garoff • Ari Helenius • Matti Sarvas • Bror Morein • Karl Matlin • Barbara Skene • Mark Marsh • Gerrit van Meer • Marja Makarow • Dorothy Bainton • Barry Gumbiner • Jitka Balcarova-Ständer • Carl-Henrik von Bonsdorff • Stephen Pfeiffer • Kalervo Metsikkö • Ivan de Curtis • Angela Wandinger-Ness • André Brändli • Robert Bacallao • Mark Bennett • Robert Parton • Morgane Bomsel • Kristian Prydz • Sanjay Pimplikar • Ian Mather • Marianne Lehnert-Wilzenski • Teymuras Kurzchalia • Jean Grünberg • Marino Zerial • Philippe Chavrier • Cecilia Bucci • Toshihide Kobayashi • Lukas Huber • Paul Dupree • Vesa Olkkonen • Klaus Fiedler • Petri Auvinen • Masayuki Murata • Elina Ikonen • Frank Lafont • Sandra Lecat • Tamotsu Yoshimori • Anna-Maria Fra • Johan Peränen • Peter Scheiffele • Daniele Zacchetti • Kwang Ho Cheong • Patrick Keller • Jürgen Benting • Anton Rietveld • Paul Verkade • Derek Toomre • Ulla Lahtinen • Michel Bagnat

DRESDEN | 2001 – 2013

Robert Ehehalt • Kim Ekroos • Joachim Füllekrug • Patrick Keller • Ulla Lahtinen • Masanori Honsho • Doris Meder • Tomasz Proszynski • Sebastian Schuck • Christer Ejsing • Robin Klemm • Lucie Kalvodova • Soazig Lelay • Sabine Buschhorn • Mathias Gerl • Otilia Viera • Vinchil Vaz • Aki Manninen • Juha Torkko • Lawrence Rajendran • Ünal Coskun • Xinwang Cao • Michal Surma • Daniel Lingwood • Hermann-Josef Kaiser • Julio Sampaio • Michal Grzybek • Ilya Levental • Adam Stefanko • Sandra Cordo • Robert Ernst • Christian Klose • Rashmi Mishra • Tobias Zahn • Cornelia Schroeder • Rashi Tiwari • Erdinc Sezgin • James Saenz

Acknowledgements

Having written a book about my life, there are so many people I would like to thank that this alone would be enough to fill a book. So I will focus on those who have been closest to me. Number one is my wife, Carola. She has been the inspiration of my life. What luck that she accepted me as her partner. She has supported me throughout our marriage and has also given important input to my book.

Our three children, Katja, Mikael, and Matias, have been our delight. What a wonder that the family remained so closely knit despite having a father that excelled in absences because of his work. Thanks for all your input. I hug you all. My thanks also go to my sister, Majlen, and her husband, Ari Helenius, who was my graduate student and postdoc, for their comments on the text. We are still good friends. With their children, Jonne and Ira, our two families spent eleven years living separately but together in Helsinki and in Heidelberg, having a great time.

With my brother, Tom, I also have a close relationship. All three families spend summer vacations together on the island Mölandet, close to Helsinki, where we enjoy the Finnish summer with its Nordic light.

This book was first published in Swedish, which is the second official language of Finland and my mother tongue, under the title *Forskningens olidliga lätthet*. Thanks to my publisher of the Swedish book, Sara Ehnholm-Hielm, and also to Hanna Ylöstalo and Helena Kajander. The photograph for the cover was taken by André Wirsig. Kristian Donner, a well-known Finnish neuroscientist, translated the Swedish text into English. Thank you; I could not have persuaded myself to do that after having written the Swedish text. It was the isolation of the COVID-19 lockdown that gave me the inspiration and possibility to concentrate fully on writing the Swedish manuscript.

Many heartfelt thanks also to Carol Featherstone who edited the English version of the text after I revised it to be more relevant to scientists.

My warmest gratitude goes to the core team of directors at the MPI-CBG in Dresden – Ivan Baines, Tony Hyman, Wieland Huttner, and Marino Zerial – who together were responsible for building this fantastic research institute. What a grand time we had together. What magic a collective can achieve by working as a team. Unity of purpose equipped us to achieve our goal, to construct an altruistic environment in competitive research. Thanks for this tremendous effort and for your comments on this text.

Finally, I would like to thank all those who have been my associates and colleagues throughout my life in research. After all these years, I have so many friends and allies around the world. Research is indeed international. Special thanks go to all those who were members of my research group. It was such a privilege to work with you all. On the following page, I have listed all the scientists who were active in my lab over the years and are now spread over the world. I especially thank James, Ilya, Robert and Ünal for their brainstorming that produced the title of this book.

About the Author

Kai Simons is not only one of the most recognised figures in modern cell biology worldwide, he has had a huge influence on research culture in Europe. In this book, he proclaims the importance of collaboration and co-operation between scientists in investing young researchers with the means to carry out original, innovative research and of creating youthful, dynamic, interactive research institutes. This belief is reflected today throughout Europe and beyond in the shape of individual careers and research institutions built on the principles he helped establish at the European Molecular Biology Laboratory in Heidelberg, the Max Planck Institute for Cell Biology and Genetics in Dresden and the European Research Council, which has become the most important agency funding excellent fundamental research in Europe.

9 781917 007023

Printed by BoD"in Norderstedt, Germany